THE TUNNEL
THE CHANNEL AND BEYOND

THE TUNNEL
THE CHANNEL AND BEYOND

Editor:
BRONWEN JONES

Contributing authors:
JOHN ARDILL, NICHOLAS COMFORT,
PETER DE IONNO, MICK HAMER,
MURRAY HUGHES, BRONWEN JONES,
TOM ROWLAND, and KATHY WATSON

ELLIS HORWOOD LIMITED
Publishers · Chichester

Halsted Press: a division of
JOHN WILEY & SONS
New York · Chichester · Brisbane · Toronto

First published in 1987 by
ELLIS HORWOOD LIMITED
Market Cross House, Cooper Street,
Chichester, West Sussex, PO19 1EB, England
The publisher's colophon is reproduced from James Gillison's drawing of the ancient Market Cross, Chichester.

Distributors:

Australia and New Zealand:
JACARANDA WILEY LIMITED
GPO Box 859, Brisbane, Queensland 4001, Australia
Canada:
JOHN WILEY & SONS CANADA LIMITED
22 Worcester Road, Rexdale, Ontario, Canada
Europe and Africa:
JOHN WILEY & SONS LIMITED
Baffins Lane, Chichester, West Sussex, England
North and South America and the rest of the world:
Halsted Press: a division of
JOHN WILEY & SONS
605 Third Avenue, New York, NY 10158, USA

British Library Cataloguing in Publication Data
The Tunnel: the Channel and beyond.
1. Tunnels — English Channel
I. Jones, Bronwen   II. Ardill, John
385'.312   HE380.G7
Library of Congress CIP Data also available
ISBN 0–7458–0277–X (Ellis Horwood Limited)
ISBN 0–470–20929–1 (Halsted Press)

Printed in Great Britain by Billing & Sons, Worcester

Table of Contents

Preface

Tunnels have fascinated me for years. Before even my first job as a journalist, on a magazine called *Tunnels & Tunnelling*, I could be found with hard hat and overalls, driving a tunnelling machine through the sandstone of Seattle in Washington State or crawling along between pit props in British coal mines. That others did not share this fascination was all too clearly illustrated by the publication's nickname, in the trade press, of 'Bores & Boring'. But the Channel Tunnel or, more fairly, the idea of constructing a link across the Channel, far from boring, has intrigued engineers, inventors, politicians and many others for over 200 years.

A large number of books have catalogued the colourful history of previous attempts. At least two other books published in 1987 deal with this topic again. But my decision, initially to write a book and then instead to co-write and edit, was from a sense of frustration at people only ever telling half of the story. A tunnel, or indeed any piece of infrastructure, cannot function alone; it is inextricably linked not just to physical structures, but to the people who design, build and eventually use it.

Civil engineering may be perceived as mud, steel and concrete but it is like blood to the body of the land.

It is not possible to cover all the areas on which the Tunnel impinges, at least not in this first edition; a short time span in which to write it and insufficient French contacts makes the book more British-oriented than I would wish. But many facets are dealt with, including finance, politics, tunnelling technology, trains and safety, environment, economics and employment in Kent and the Nord-Pas de Calais, history, ports and shipping, and a look at future links in the world. My seven co-authors were selected primarily for their proven interest in the Tunnel and some for their specialised knowledge. Most of them write regularly for daily newspapers, all I hope have retained their individual catch-the-deadline style, essential to describe a project waiting impatiently to break, alive and kicking, into the world.

In February 1987, I and eleven colleagues were made redundant when the magazines on which we worked fell prey to a takeover bid. This gave a strong impetus to the book and also meant that my former secretary, Rachel Henshaw, could join 'The Tunnel' team as researcher and aide-de-camp. Her unstinting help and enthusiasm cannot be repaid. My grateful thanks must also go to my aunt, the Baroness White, for her continual advice and particularly for editorial comment on political and environmental content. Many others, too numerous to mention, helped provide photographs and information, although in my enforced unemployment I enjoyed visiting sites affected by the project and most of the 'on location' shots included herein are my own.

The unsung heroes of the Tunnel are its engineers and designers. Although many nationalities have contributed, France and Britain should be proud of their skilled team forging the link. My particular thanks are extended to UK consulting engineers Mott Hay and Anderson for technical details supplied.

The book is the only one to approach the Channel Tunnel project in such a comprehensive way. A multi-functional text, the aim is for people with specialist interest to read first 'their' chapter and then to delve into all the other aspects of the project. Others who would like a gentle grounding in this earthy subject, will find it factual but not filled with daunting technical vocabulary. And for the school or university student, the different sections provide information and analysis relevant to politics, ecology, civil engineering, economics, or town and country planning courses.

It is vital reading for those intending to buy shares in the Equity 3 placing in Autumn 1987, and indeed for any who wish to be well informed on this interesting area of current affairs, that will remain current for at least another six years.

May 1987   BRONWEN JONES

Introduction

The British are very good at talking; sometimes I think they can do little else. The French, as chauvinistic and arrogant as the British, talk too. The difference is: they seem to talk about what they have done, we talk about what we might do.

People have talked about the Channel Tunnel for well over 200 years. Very little extra information has been added to the general debate in the last 100 years. Anyone who has researched the history of the project will understand the frustration welling up inside as each "new" idea — whether for or against — is mentioned. It has all been said before. And all who have been involved in promoting European unity just wish that one of a vast number of adages applicable to the Tunnel would be brought into play - More action, less words! Here are more words but they are written in the hope that the projects' eventual completion can now be seen as a *fait accompli*.

The writers of this book essentially believe that a physical link between the UK and the rest of Europe is a good idea. Most of them do not care whether or not it is the scheme devised by Eurotunnel, but as this appears to be the closest to fruition of any scheme suggested in the last two centuries, I at least, would like to hasten its progress.

But, no one can ignore the tragic losses that will result along the way. No one who is not involved commercially can dare to deny the environmental damage or the short-term localised negative effects on employment. However, if one stands neither to gain nor to lose money from the Tunnel venture, as is currently the case with all the authors writing herein, then a realistic perspective evolves. And the consensus is: there are far more positive points about the Tunnel than there are negative points. This reflects both Governments' attitude, when they gave a joint political commitment to the idea of a fixed link in 1985.

The authors range in age from 29 to 42; all but one are British; many have children. They live or work in the London area; some by birthright are only too aware of the north–south divide. All have witnessed the declining aspirations of young Britons in recent years and see a more European outlook as a potential lifeline for the future prosperity of the UK.

I cannot comment further on the perspective of others; I will try to explain why I see this physical link to be both so important and yet so insignificant. The Tunnel will just be a relatively short stretch of railway track, just part of a transport system. It has generated unending debate, it is technically interesting but there are no space-age concepts, nothing is impossible either in the digging, the politics or the finance. The logistics and the methodology are difficult but it is all feasible.

What is more interesting is the effect of this Tunnel, in time and geographically. Imagine the Tunnel to be like an explosion on time-delay sequence film. The impulse in the centre radiates outwards — in fragments, heat, light — theoretically never stopping. The Tunnel will implode, pulling in people and freight and then explode until borders become insignificant, effects pulsing out from the epicentre into an ever-increasing hinterland.

I see, a nation that once stood proud, now clinging to memories. There is an opportunity to benefit from our experience if we recognise that we are too small to stand alone. Our individual greatness is in the past — and was exacted at too high a cost in the days of Empire. Now, if the UK would take projects like the Tunnel as a catalyst — not a curer of all ills, and temper the reaction into a form which different areas require, there is a chance for a change in attitude and thence a change in fortune. Such ideals do not have to be along either capitalist or socialist lines to work.

In researching this book, several facets of the current British character have been brought home to me, and were an unpleasant shock. Xenophobia, extreme selfishness, hypocrisy, greed, short-sightedness, and, most of all, any perspective of fairness. There was a deeply-held conviction that the UK was having a particularly hard time and, whatever the dispute, whether between nations or companies, there appeared to be absolutely no compassion for "the other side". I am sure other nations have these problems as well, but the depth of insularity, ostrich-head-in-the-sand pride and agoraphobia preventing us from looking over the protective moat made me wonder why the whole race, lemming-like did not seal its fate and exit over Shakespeare Cliff.

In the Channel Tunnel debate complaints about economic incentives in northern France were easy to make. The amount of energy that went into compiling all the research, leaflets, publicity, could have been usefully spent devising a different strategy for the UK, to provide competitive incentives (but with more imagination than tit-for-tat tactics) for European or other companies to locate in southern Britain.

Or, perhaps more important, the shipping unions, while looking after their members' interests in the Channel and neighbouring ports could also have taken an active and enthusiastic role in developing or redeveloping west coast ports for the challenges that they can offer in six years time for trans-Atlantic trade. These things will not happen without industry, government and unions working together towards objectives they profess to have in common, namely employment and flourishing industry.

Authors on location. Five of the writers went to look at relevant sites in Kent. Here (from left to right) Peter De Ionno, Bronwen Jones, Tom Rowland, Nicholas Comfort and Kathy Watson stand in front of the Whitaker tunnelling machine, abandoned in the cliffs of Folkestone after failing to gain support for another Chunnelling venture in the 1920s.

And other unions could work with the Confederation of British Industry, both to help companies expand to take on the challenge of Tunnel-related contracts and at the same time to ensure that this extra capacity is utilised once the Tunnel is operational, with aggressive 'into Europe' marketing rather than redundancies when major contracts are finished. It is worth gearing up for a large order if good management are taken on to determine and exploit further opportunities. When Eurotunnel's contractors, Transmanche Link, asked companies to tender for high quality (spheroidal graphite) cast iron tunnel lining segments, it found no European company could take on the volume of work at a realistic price. Smaller tenders were issued instead. Why did British industry not step in to accept the challenge? Admittedly, one British company particularly well-suited to the task is French-owned!

And in the lobbying, how often did environmental groups remind people about the number of juggernauts that would no longer thunder through small villages? Why did the government, after actively support-ing a Tunnel in principle, allow eight further freightliner depots to close, making long-term, long-distance freight haulage through the Tunnel of decreased benefit to certain regions? How could a company that had proposed to build a fixed link that would have generated more chalk spoil than any of its competitors, and been more dangerous to travel in than most of the other schemes suggested, then take an extremely active role in opposing particularly these two major issues? Was it just sour grapes?

Unscrupulous self-centred campaigning is not uncommon on other issues, but I felt it reached a peak for the Tunnel. Parliament watchers commented cynically that so-and-so said that because he is paid by the ferry companies, and another one said that because he used to work for the railways, and X cannot be on a committee because he wants the Tunnel but owing to the location of his constituency he could not possibly say so. Friends (or paid consultants) of pro- and anti-Tunnel groups provided lists of their colleagues to be lobbied on specific topics, and even members of a select committee were approached while their work was in progress by an organisation that knew this to be unethical. It seemed no holds were barred and those on the periphery were not interested in the antics of those in the fray.

The House of Lords Select Committee fortunately seemed particularly astute at recognising some of the bare-knuckle behaviour and differen-tiating between blatant propaganda and solid fact. Sometimes the media was less than fair in its reporting. Most MPs seemed unable to look beyond the looming election (even in early 1986) and had insufficient political maturity to contemplate implications for UK trade a decade or two ahead. It is a pity that more Euro MPs (MEPs) were not involved in drawing up the Bill.

Current cross-Channel ferry economics strongly favour the British. If the opposite applied, no doubt a protest body would have been set up by now to complain about it. When the Eurotunnel idea of 50:50 Franco-British ownership was proposed, people complained about altering the imbalance; obviously the long-suffering French did not. A host of anecdotes on the mean-minded, narrow-minded, attitudes of so many involved in the project could fill a book in their own right.

When Dover Harbour Board suggested that construction of a sea wall or the testing prior to construction could locate shipwrecks with explosives on board, it was not issuing a kind warning on behalf of the drilling crews, merely seeking to promote its duty to protect mariners to whom it was answerable. To do both would have been more acceptable.

Eurotunnel has some wealthy backers and, undoubtedly, if it is allowed to build and operate its Tunnel, will become an exceptionally wealthy company. At the time of petitioning Parliament, however, its assets were very definitely finite. To smooth a path through the almost-fashionable opposition, the company made offers well in excess of legal requirements for compensation on compulsory purchase of property or other disruptive activities. Whether this was inherent generosity or calculated oiling of the political wheels, is in many ways immaterial. The result was that vast numbers of people who could, or could not, prove they would have a dustier, noisier, lifestyle demanded double glazing. Those who thought they might have a construction lorry pass within a kilometre of their home asked for new roads to be built. Those who would not be affected by the scheme demanded that they were. Hang-gliders complained about loss of land that they did not actually have the right to use in the first place. Eveyone expected Eurotunnel to pay from what they obviously perceived to be overflowing coffers. Compensation is essential. Parity of treatment, likewise. But the level of greed of those who could not seriously justify their claims was astounding.

The genuine cases were often very sad. Beautiful houses that people had retired to and worked hard on, would be demolished. The decision to destroy five houses in the village of Newington was a local choice of the lesser of two evils for an access route to the new terminal at Cheriton. A listed building that will be lost, already stands on the edge of the motorway... the sheep in the garden did not seem bothered by the cars, so perhaps they at least will live happily alongside the busy terminal. A counter argument is that property on the French or British coasts, although no longer in such an "idyllic" location will increase dramatically in value due to its proximity to either terminal.

Whatever happens there have to be compromises. If there is no environmental disruption there are no extra jobs and if traffic is directed immediately towards Paris or London, local communities keep more green grass but get no extra trade. A balance has to be struck. As to Eurotunnel paying for all the infrastructure from which it benefits, this is perfectly acceptable if the same applies to all sea- and air-ports. The more normal method of payment is through rates. Likewise for services such as police and firemen. When the *Herald of Free Enterprise* tragedy occurred early in 1987 and the vessel capsized off the Belgian coast, it would have been unthinkable to demand that the shipping companies pay extra for the assistance of the British services involved. All ports have extra security staff employed, but it did appear that Eurotunnel was being asked to pay above the normal odds to cover any possible disaster, private company or not.

Murray Hughes, railway specialist, went to see the 55-metre diameter shaft under construction in Fréthun, in March 1987.

As to xenophobia, well, the French do not appear to be particularly nervous of the British although they do not necessarily hold them in high regard. Twice, the British have unilaterally reneged on Channel Tunnel understandings, in the 1970s and in the 1880s. But the British attitude to the French is odd to say the least. Curiously, a middle-ranking employee of TML once even told me that he did not like the French! Hardly a positive attitude for working in a bi-national company.

The more popular stereotypes say: "Frenchmen are short with navy and white horizontally-striped shirts, flat dark berets, tanned skin, dark hair probably with a moustache, and they wear baggy trousers. They carry onions over one shoulder as they pedal along on bicycles and have a very long thin loaf of bread tucked under the other arm. They eat strange things like frogs, snails and lots of garlic — and enjoy them.

"The lesser noticed Frenchman wears extremely well-tailored clothes, almost to the extent of being pretty. He croons love songs with a delightful accent and typically has a name with a very soft "sh" sound in it. Adored by British women, he is disliked intensely by British men. And as to French women. They either wear feathers on their heads and have full skirts with petticoats exhibited by kicking their legs in the air. Or they have a certain je ne sais quoi and are alarmingly neatly dressed in crisp suits that never crease. The British man is terrified of them; even the topless bronzed femme fatale on the beach proves unnerving rather than inviting."

What do the French think of the British? "Well", said one, "we do not think of the British very often anyway, we see no good reason to." The British are quite useful to export lamb, apples, wine and nuclear electricity to, but when it comes to le crunch, the offshore island does almost nothing but complain or try to alter perfectly acceptable agricultural policies that have worked well for years. Also, it joins inter-European scientific projects and then seems unable to pay its way.

"The British man is not only cold and unromantic, but he does not like to wash and obviously does not care what his clothes look like. The young men are violent thugs, most dangerous at sporting events overseas. Old men are laughably quaint which is why they cannot do business — look at the trouble your "City" had raising the Equity 2 money.

"British girls [usually British is taken as synonymous with English] have often got beautiful skin and are easy to impress. They are not sophisticated in manner or appearance. Britain is plagued by strikes,'green' groups and nuclear protesters — like a permanent population of students." Prejudice works both ways.

Interestingly the anti-Tunnel groups have been determined not just to stress the damage the Tunnel could do to southern Britain and elsewhere in the country, but also found it necessary to depict northern France as an industrial quagmire, almost a place with no 'environment' left to protect. Kent is called the garden of England. The French describe the Nord–Pas de Calais region as the gateway to Europe, one tourist brochure commenting "..and the river Authie which threads its way through a string of villages. Of all the areas …this is where the expression 'douce France' (gentle France) rings true".

Tourism

The Nord and Pas de Calais regions combine to form an area of similar size to Northern Ireland. It has four million inhabitants, forming over seven per cent of the total French population. Both northern France and southern Britain will gain from tourism generated by the Tunnel. Kent has sandy beaches, rocky shores, hills, woods, pretty villages and ancient towns. So too, has the Nord–Pas de Calais, with its varied topography and vegetation related to the geology beneath.

Calais, the medieval city of St Omer and Dunkerque dominate the coastal plains. They are linked by the river Aa, which marks part of the historical West Flanders boundary. Long sandy duned beaches are fringed with fields of corn, flax and intermingled wild flowers on reclaimed land. And although a quarter of the land is occupied by ports and associated industries, there are still a multitude of market gardens hemmed by canals with flat barges carrying produce to market.

Boulogne, at the northern tip of the region, provides a view of the white cliffs of Dover with Englishmen gazing back at the white cliffs of France. More sand dunes and the picturesque headlands of Alprech, Gris-Nez and Blanc-Nez (white nose), lead inland to pasture outlined by hedgerows, and forest. The more hilly Houtland is predominantly farming country, scattered with churches and windmills. In the towns of Bergues, Hazebrouck and Bailleul, local people often speak Flemish, reflecting the fluctuating borderline over the centuries. And on the plains of Lys, towns sit by the river responsible for their growth. Not so long ago, linen used to be bleached on the river banks. In Pevele country though, land is flat, green, full of small streams overhung by weeping willows. Fields of sugar beet supply the local sugar industry. And hunting-fishing country is to be found in the Valley of the Sensee in the marshlands crowded with waterfowl near Douai.

Vast cemeteries in the Artois region prove the clear but sad link with Britain, forged in two world wars. Peaceful now, they bring many visitors who want to remember. Woodland on chalky soils, coal mining relics on the landscape, patchwork fields, herons in the marshes; some parts could almost be Kent but the flavour is very definitely French.

The potentially prosperous industrial base of the area is not to be denied though, with half a million people living in Lille, Villeneuve-d'Ascq, Roubaix and Tourcoing. Here it is hoped the region can halt the industrial decline of recent years and the Channel Tunnel is seen as a vital contributing factor in all future equations. A local brochure states "…we intend to take full advantage of the privileged geographical situation we possess. At the centre of a triangle made up of London, Brussels and Paris and which the Channel Tunnel and North European Fast Passenger Train will eventually turn into a nerve centre". It would have to be an extreme pessimist to say that Kent, a mere 40 kilometres away from Nord–Pas de Calais, is going to be a nerveless centre.

But Britain feels more secure in pessimism. If expectations are nil, then none can complain at lack of achievement. Yet Kent can attract business, and tourism with careful planning need not damage its admittedly precarious environment. The orchards, oast houses and hop fields are still there. But ugly industrial estates and shabby seaside towns are present too. Since my village childhood in Kent, the multitude of wild flowers that laced the verges and hedgerows in a gorgeous spectrum have dwindled to a mere handful of purple vetch, golden buttercups, cow parsley and occasional white campion. Yes, I can understand the environmentalists' fears.

John Ardill, environmental author in *The Guardian* offices in London.

What is there to tempt the tourist in either land in 1993? Much as now, unless companies use the intervening years to prepare for what many consider a welcome invasion. Of course, tourists do not just have to consider the areas adjacent to each end of the Tunnel, but can also look forward to easier travelling beyond.

In 1987, a determined and relatively wealthy person working in London, could leave the city early on Friday evening, and barring delays on the underground train or at the airport, could hope to arrive in the skiing resort of Mottaret in France by 11pm. Enjoying the 190 linked ski lifts, and with 500 snow cannons guaranteeing artificial blizzards if weather conditions are not sufficiently 'good', would cost about £250 with food and ski hire included. The return trip would have to start at 3.30pm on Sunday. In 1993, a train service will probably allow a similar journey at far less cost, and quite likely in less time. There will be no need to return home early but an overnight sleeper train will bring the traveller refreshed and suntanned back into the centre of London, ready for work without Monday morning blues.

The possibilities of frontier-less thinking are endless. I went to school in Canterbury in Kent in the 1970s and set off to London to find work at the age of eighteen. In the mid-1990s the opportunities might look just as good in Lille and travelling would be equally easy, even commuting each way will be feasible. There may be a Europe without internal borders by then and perhaps Britons will be proud to be European. With potentially all Europe's countries as members of the Community, the four super-powers, China, The Soviet Union, the USA and Europe will hold discussions about the rising strength of Africa and the effects of greater Asian unity.

In 1999... with the Tunnel already proving under capacity, the ferries lapping up the extra trade and high speed cross-Channel catamarans taking pleasure trippers rather than the dedicated businessmen they were intended for, the magnificent spans of Chanbridge's aerodynamic glass traffic tubes are being linked 30 metres above sea-level, the carbon fibre masts of sailing ships passing a mere five metres beneath them.

Travellers insert their passport-EFTPOS (electronic funds transfer at point of sale) cards into machines at the single checkpoint for each direction of travel and press a fingerprint card-verification button. The random-number computer selection instruct cars occasionally to divert into another compound for customs checking, and the laser-analyser reads number plates, directing any suspect vehicles to one side. Smuggled pets no longer a problem, with rabies almost eradicated from Europe in any case and the infra-red locators in France preventing any vehicle with non-human passengers from continuing its journey into the UK.

Meanwhile, in the Ashford International Conference Centre, five European telecommunications companies haggled over the rent Chanbridge want to charge to carry their fibre optic links beneath the carriageways. The new laser amplifier repeaters every 250 kilometres have been installed across Britain, France and Spain, but the main link stops short of the floating tunnel across the Strait of Gibraltar, as Euro-T, major shareholder, waits to see what rate Chanbridge sets for carrying the service.

And at Cheriton, all shuttles are suddenly stopped to let the Canterbury–Lille ambulance train through. A few of the waiting passengers wander over to the supermarket siding, muttering that it really is not worth the effort just for chocolates and perfume. It was more than ten years since the EEC had stopped duty free sales of alcohol and tobacco on health grounds.

Back to 1987. As the book went to press in early May, prospects looked good. The Bill authorising construction of the Tunnel was well on its way through Parliament when Mrs Thatcher called an election for June 11. In just over a year it had got through the House of Commons and survived 66 days of scrutiny by Select Committees of both the Commons and the House of Lords, who took evidence from hundreds of objectors. While the two committees proposed 179 amendments between them, none struck at the fundamentals of the project. Barring the unexpected election of a hostile government, the Bill appeared on course for Royal Assent towards the end of July 1987. As the Bill progressed through the House of Lords, parallel measures made speedier headway in the French National Assembly.

Of 109 amendments published on May 7, the changes covered included a multitude of minor road works in addition to the plans for the A20 and the M20 motorway, restriction details on chalk spoil dumping, protection of navigation from any interference during construction of the spoil disposal platform and a new relief sewer for the town of Folkestone to cater also for terminal site drainage.

Flexilink, the vociferous anti-Tunnel group was quiet on safety issues. It did not seem appropriate while bodies from the *Herald of Free Enterprise* were still being washed up on the Belgian coast. Flexilink's case was not helped by a collision between 5590-tonne Sealink car ferry *Hengist* and a fishing trawler *La Glorieuse Vierge Marie II* in Boulogne, in which the smaller vessel sank and three men died. The media keenly watched for further ship troubles to add to coverage from Belgium and reported almost instantly a collision between two Sealink ferries on May 1, in the fog off Dover. The ferry *Cambridge* retired to the port with a crumpled bow almost to the waterline while the 4648-tonne *St Eloi* nursed a large hole on its starboard side, and also docked at Dover with assistance from tugs. Luckily no one was hurt.

One member company of Flexilink, Hoverspeed, started a retaliatory advertising campaign on the theme of Tunnel Vision in May. A large black rectangle represented the Tunnel and underneath it said, "If you want to wait 6 years you might be able to tunnel your way to France in 35 minutes. What do people see in it?". The lower half of the page had a hovercraft gliding on spray with a perfect sky behind it. The caption beneath read "Alternatively you can take the flyover today in just 35 minutes. And enjoy the view." It is only a pity that hovercraft and jetfoil systems are so limited by the weather, and not just in the winter months. When both systems abandoned services on the windy UK bank holiday on Monday May 5, 1987, customers might have liked the maligned tunnelling option to be available.

Undaunted by accidents or weather, Flexilink prepared for its major onslaught on Equity 3 expected in the autumn and turned to a new line of attack, commissioning transport economist Mr Stephen Plowden to report on the traffic figures on which Eurotunnel based its forecasts. Plowden said that "both traffic and revenue forecasts were unreliable". Flexilink and Sealink are seen by many to be synonymous and in this context it is interesting to note a comment made by Lord Ampthill, Chairman of the House of Lords Select Committee on the Channel Tunnel, on April 23, 1987. He said to a witness, "I think that you will accept that Sealink has been making a vast amount of propaganda against the Tunnel, in its own interests, and that it is perfectly entitled to do, but I would personally not take the accuracy of some of the statements made

Mick Hamer works in an author's paradise of curious cats, grand piano, ancient black and gold manual typewriters, and no one to complain about the muddle. April 1987.

too highly." On the same day, the highly respected Lord said, "It is my personal view that the ferries will be around for a great many decades...."

Meanwhile, the Council for the Protection of Rural England was furiously up in arms. At a conference on May 11, UK Environment Minister William Waldegrave stepped into fierce criticism over Government handling of the Channel Tunnel Bill. Mr Robin Grove-White, Director of CPRE said, "Much of Parliament's scrutiny of the implications of the Bill has been a charade... Last year, ministers promised repeatedly that all objectors would be given a fair hearing — an important promise given the range of impacts (environmental and otherwise) expected from the Tunnel, Europe's largest-ever private sector engineering development. Those assurances have proved empty."

He went on to say, "it's all the more disturbing that we now hear strong rumours in Whitehall and Westminster of more, similar Bills on the way. Both British Coal and the Central Electricity Generating Board, it is rumoured strongly, are looking with interest at the possible use of Private Bill procedure to force through unpopular new coal mines or nuclear power station proposals, without public inquiries. That would be calamitous if it proved true."

All opposition to the Channel Tunnel Bill, whether in organised groups or from concerned individuals, raised vital issues and achieved some important amendments. The tactics of those who were opposed to the Bill in principle, in seeking merely to delay and frustrate legislation, may have taken valuable time from those who reluctantly accepted the inevitable and sought to achieve useful compromises. No doubt debate will continue until construction work on the second fixed link starts. Even if there were no genuine complaints (in 1987 there certainly were important issues to be resolved), old habits die hard, and in the intermingling of European cultures that will now follow it would be sad to lose the dedicated British protester.

In preparing this book, I invited a variety of individuals or groups to contribute material to put their unedited perspective on record. Many, including Eurotunnel, did not take up the offer, for reasons of time, or for no given reason at all.

There follow four contributions. The first was written by David Puttnam, President of the Council for the Protection of Rural England and international film producer. There follows a contribution from Ms Maureen Tomison, a director of Flexilink. Mr Michael Diacono deputy managing director of the Sally Line gives his view and Lord Underhill brings the section to a close, as deputy leader of the opposition in the House of Lords. It should be noted that Lord Underhill's contribution was written in the early stages of the Lords Select Committee and some of the issues have been in part, resolved. BRONWEN JONES

Council for the Protection of Rural England: Environmental controversy and the Channel Tunnel

The Channel Tunnel scheme has aroused enormous controversy because of its likely environmental impacts. Not only will the Tunnel's terminal and its associated works cause immense upheaval to attractive countryside around Folkestone, but the scheme will also be a huge generator of traffic affecting many communities and a likely catalyst for other, secondary developments throughout Kent bringing major and permanent change to the Garden of England.

The Council for the Protection of Rural England (CPRE) has led much of the campaign against many of these impacts. Both before and during the passage of the Channel Tunnel Bill through Parliament we have fought hard for amendments to the scheme, to protect Kent. But the Government has remained largely obdurate, its thinking dominated overwhelmingly by the imperative of getting the Bill through Parliament in the shortest possible time.

In January 1986, CPRE helped form the Kent Action Group — a consortium of country-wide organisations including the Kent Trust for Nature Conservation (KTNC), Kent Federation of Amenity Societies (KFAS), Weald of Kent Preservation Society (WKPS), Kent Association of

Local Councils (KALC), Kent Ramblers, Association and CPRE Kent. Several of these bodies (though not CPRE itself) had initially been opposed root and branch to the Tunnel, because of the upheaval it would cause to Kent. But all of them recognised rapidly that, with a Bill in prospect, it was vital to combine forces pragmatically and seek changes to contain its impacts, by representations to the Parliamentary committees scrutinising the Bill.

The Action Group's members shared several crucial concerns, all major. The 350-acre terminal at Cheriton and the network of approach roads leading to and around Newington would devastate a hitherto undisturbed area. Traffic impacts on nearby villages and country roads would be extensive. Road movements of millions of tonnes of construction materials would aggravate the pressures on local roads and communities. And disposal of the millions of tonnes of spoil excavated from the Tunnel itself, would cause a major blot at the foot of famous Shakespeare Cliff.

But these were simply local impacts. Quite as disturbing were the prospects of wide-ranging development up and down Kent, following construction of the Tunnel — new warehousing, retailing and tourism centres, and industrial and housing developments attracted by the magnetic pull of the Tunnel within the region. Controversy about this dimension of the scheme — barely hinted at in Eurotunnel's own Environmental Impact Analysis — grew steadily during the Bill's Parliamentary stages, fuelled by the publication of a specialist consultants' report commissioned by CPRE and KTNC (with World Wildlife Fund (UK)'s help) and other studies by Kent local authorities.

And then there was the A20. When the Channel Tunnel Bill was published in April 1986, the Action Group had been astonished to find it provided not only for the Tunnel and all its works, but also for a new stretch of motorway from Folkestone to Dover, to be pushed slap through the magnificent countryside of the Kent Area of Outstanding Natural Beauty. Normally, such a scheme — particularly one affecting such sensitive a landscape and bearing no direct relation to the Tunnel proposal itself — would be the subject of a local public inquiry. But not in this case. It became clear that the Government's inclusion of the scheme in the Bill was simply an opportunistic move to avoid the scrutiny that would have resulted from such a public inquiry. CPRE and its fellow objectors fought to have the road struck from the Bill, or, at the very least, to have the route altered. The Countryside Commission too, the Government's own statutory landscape watchdog, were strongly critical of the road.

These then were the key issues on which the member bodies of the Kent Action Group petitioned Parliament during the various Committee

Much of the Nord-Pas de Calais contains beautiful Gothic-style architecture.

stages of the Channel Tunnel Bill during 1986 and 1987. The Action Group was far from alone. A record number of other outside individuals and organisations also raised formal objections — 4900 in the Commons and 1400 in the Lords. Kent local authorities, including Kent County Council, also petitioned vigorously.

The results of this ferment of activity were mixed. There were a number of changes to the Bill, the most important being new arrangements to restrict the impact of the roads into and out of the Terminal itself (proposed Shepway District Council) and constraints on lorry movements of construction materials in the country. But on other central issues, the Government made few concessions, to the dismay of many people and organisations.

This failure by the Government had its own wider implications. From the outset of the current Channel Tunnel scheme in October 1985, the

Government's emphasis had been on speed at all costs. A public inquiry was ruled out from the start by Transport Secretary Nicholas Ridley, and environmental scrutiny was treated as less than central. True, all the initial rival contenders for the concession to build a fixed link were obliged to demonstrate awareness of the range of environmental implications of their proposals. And true, Eurotunnel's scheme was probably the best considered (the 'least bad') of the schemes on offer. Nevertheless, once Eurotunnel's scheme had been accepted, the timetable imposed by the Government from inception to Royal Assent of the Bill was grotesquely tight for a scheme of this magnitude. This led CPRE and others to press the Government repeatedly (and largely fruitlessly) for more time and care. MPs Jonathan Aitken and Nick Raynsford backed up these plans with ingenious tactics of their own.

David Puttnam, Council for the Protection of Rural England President and international film producer.

Nevertheless, serious scrutiny of the Tunnel's impacts and if possible measures to alleviate them was inadequate. The Select Committees in both the Commons and the Lords compressed thousands of individual objections into a highly restricted timetable, creating an alarming precedent for cavalier treatment of public concerns in similar future cases.

Advocates of the Tunnel have argued that too many environmental safeguards in Kent could lead to the associated developments — and hence economic benefits — being located across the Channel in the Pas de Calais, where the French would welcome them, less discriminatingly perhaps, with open arms. This dilemma is real. The inescapable fact is that Kent is a beautiful, crowded and relatively prosperous part of Britian, whilst the Pas de Calais is an emptier and more deprived (though geographically highly convenient) region. Especially strong and sensitive planning restraint and planning guidance are a *sine qua non* in Kent if any benefits of the Tunnel are to be gained without the corresponding disadvantages.

The perpetual concern of CPRE and the other key environmental objectors to the Channel Tunnel Bill has been that the Government has not faced up to this fact adequately.

That is why, despite the Government's reluctance to concede major changes to the Bill, — and despite the welcome arousals of Kent County Council and the various district councils that they will protect Kent's countryside — we have continued to the bitter end to campaign for a sounder approach. The national and local environmental policies that are applied once construction begins will have continuing importance for this wonderful corner of England, if it is not to suffer irreparable damage.

Flexilink floats its view

When Mrs Thatcher and President Mitterrand announced at the end of 1984, that they wished to see some form of fixed link built across the Channel, it was for the existing cross-Channel operators an all-too-familiar situation: normally a fiercely competitive group, they had been required to co-operate in the past when other schemes had been proposed, in order to pool their unique knowledge of a highly competitive market.

On this occasion there were three factors that made the announcement exceptional: the political impetus behind the project given the support of a determined Conservative Prime Minister and a socialist President; their resolve that the private sector should pay for the scheme; the fact that there was no one scheme under consideration. The two Governments had decided to set the criteria under which interested parties would submit their schemes and from those a "winner" would be selected.

When representatives of the Dover Harbour Board, Townsend Thoresen, Sealink and the Port of Calais — the founder members of Flexilink — met to consider the implications of the announcement they found themselves in some difficulties. There was no clear target in view, merely a concept, and yet that concept had to be treated seriously. In those circumstances the founders of Flexilink decided that pending the selection of a particular bridge, tunnel or brunnel they had little choice but to highlight the virtues of the existing services — for example their ability to develop and evolve with the demands of the market; to examine the implications for choice in the longer term that a fixed link presented to seek to ensure that a fixed link if built would compete on the basis of Flexilink's membership.

All this changed at the end of January 1986 when the rail-only scheme prepared by the Channel Tunnel Group, later renamed Eurotunnel, was chosen. It then became possible to concentrate on assessing the scheme's financial viability, its impact on the environment and the likely safety levels. Safety was a particular concern in respect of the shuttle scheme by means of which road vehicles were to be transported through the tunnel in enclosed wagons.

It was at this point that the balance of advantage in the debate about the future of the cross-Channel services changed sharply in favour of Flexilink. While there was no one scheme under the microscope of public attention, merely several well-financed contenders seeking to become the two Governments' final choice, all media attention was inevitably focussed on the speculation about the likely winner of such an exciting race. Yet for the 'winner' of that race, despite their vast resources, the battle had only just begun.

The British government had made it clear from the outset that there would be no public inquiry. Instead they claimed that the Hybrid Bill procedure, which would be used to enact the legislation necessary to build the Tunnel, would provide equally effective protection for the interests of those affected. The decision proved the first of many banana skins for the Government and Eurotunnel alike.

Because of an administrative oversight the Government failed to observe the proper rules for the introduction of the Channel Tunnel Bill and very nearly lost it completely at the outset. Thereafter, despite assurances from the Secretary of State for Transport that every petitioner would be given ample opportunity to be heard by Select Committees in both the Commons and the Lords, the Government found themselves overwhelmed by petitioners from Kent and elsewhere in the UK. There were even petitioners from France. In the Commons alone 4854 petitions were lodged, an all-time record.

Flexilink was accused by some critics of orchestrating a campaign against the Bill. The truth was rather different: over 4500 petitioners can not be invented.

The members of Flexilink adopted a robust stance: they would petition themselves and at the same time assist others, seafarers, local residents or Kentish consumer groups who asked for assistance. For many people the business of preparing a complex, legalistic document and then being required to appear before a group of MPs or Peers was a difficult and intimidating task. It is a measure of the depth of feeling against the Eurotunnel project that so many were prepared to do so.

The Tunnel is so much more than just a transport system. Its effects will explode through geography and time.

Many of those people, unversed in the ways of Parliament, felt that their efforts were wasted. The contrary was the case. From the day of their joint announcement the British and French governments had set an unprecedentedly tight timetable for such a huge project. The protracted discussions of the Select Committees, and also of the Commons Standing Committee, ensured that many of the flaws in the project were highlighted in a fashion that only mature consideration can achieve.

The dubious safety standards of the shuttle train were forcefully questioned by Labour MP Nick Raynsford, a member of both Select and Standing Committees. Cogent evidence from the existing cross-Channel operators led the Commons Select Committee to tighten up provisions on the Bill for fair competition. And while this wholly welcome and proper debate was going on, Eurotunnel was demonstrating to the world just how ill-prepared they were to see the scheme through: advisers came and advisers went; chairmen were turned over rapidly as well. One, Sir Nicholas Henderson, went so far as to publish a book in which he reported that Eurotunnel's own merchant bankers had believed it impossible to finance the scheme.

For Flexilink the task was straightforward. The key issues such as financing, competition, traffic forecasting and safety, were made the object of separate studies and campaigns. Other organisations such as the Fire Brigades Union were drawn in. Eventually the strength of the Flexilink case began to tell against the sheer mass of Eurotunnel's poorly marshalled forces.

Sally Line in beleaguered Thanet

In my mind there is still a big question mark over whether or not the Channel Tunnel will ever become a reality. Although politicians on both sides of the Channel maintain their commitment to it, the financial scale of the project, and the ability of Eurotunnel to raise the necessary cash to fund the next stage of its development, must cast serious doubts on its viability.

While the British government has said that tax-payers' money will not be used to prop up the Channel Tunnel, the financial commitment which Eurotunnel is demanding of British Rail and SNCF is seen by many as an indirect 'subsidy by guarantee'.

Whether or not the Channel Tunnel project goes ahead, it has acted as a catalyst to the ferry operators in giving clear focus to the challenges they face during the remainder of this century. When Sally Line entered the cross-Channel market almost six years ago, it brought fresh ideas and new standards of service which helped it carve a place in the market. Since that time, the Channel scene has altered radically. The other operators have been forced to raise their own standards and several modern new ships have already been introduced or will be soon.

The established operators have learned from Sally Line's success and have woken up to the fact that there is a vast potential to increase revenue by providing more retail and other facilities on their vessels, such as duty-free supermarkets.

The ferry industry is in a state of flux, with companies changing hands and altering their outlook. The acquisition by the P & O Group of Townsend Thoresen, and James Sherwood's takeover of Sealink, now called British Ferries, both point to a change in emphasis away from mere transportation and towards leisure-oriented travel. The introduction of 'jumbo ferries' is seen by some industry commentators as the answer to all the operators' problems. But it is never as simple as that. Certainly these vessels will help reduce costs. But until the major operators renegotiate their manning levels with the unions, they will not be able to match the competitive edge of smaller, leaner companies like Sally.

I believe ferry operators will have to fight hard to establish new markets… and keep them. There will almost certainly be a reduction in

Michael Diacono, deputy managing director of the Sally Line, lives in an unemployment blackspot, the Isle of Thanet.

fleets. For example, six jumbos operated by Townsend and British Ferries is all that is really needed to service the Calais route. Already British Ferries is calling for a form of cartel, which would certainly not be in the best interests of the travelling public. Thankfully, the Office of Fair Trading is firmly opposed to such a move.

If the major operators can achieve the sort of streamlined operation Sally has built, they will be able to face the challenges of the future — with or wihout a Channel Tunnel. But the secret of survival is financial stability and continued investment in developing new markets. The innovator will become the market leader. For example, one project in the pipeline is the development of a major leisure and recreation centre in Dunkirk, which would provide a further reason for the public to travel on Sally Line's service from Ramsgate.

Sally is working with local authorities on both sides of the Channel to develop mini-break and weekend packages to provide yet another reason why the public would want to use the ferries. The ferry operators must become marketing led, like the major tour operators, if they are to prosper in the years ahead. A ferry operator who believes that price war tactics will lead to profitable market growth is playing Russian roulette with a bullet in every chamber. All that price-cutting does is introduce red ink on the profit and loss account.

What is needed now is a steadying of the market through the introduction of realistic rates tied to a high level of service and facilities and innovative marketing. If the operators adopt this policy, they can view the future with a reasonable degree of confidence, despite the spectre of the Channel Tunnel. If instead they go for a short-term advantage by cutting rates and diluting the profits they need to invest in the future, they could be playing right into the hands of the operators of a Channel Tunnel. I cannot believe that Sally's competitors would adopt such a short-sighted policy. Any more than I can believe that City financiers will ultimately be prepared to sink their funds into a hole below the Channel.

View from the left of the Lords

My inclination was to give support to a Channel fixed link, but only if any project was to include a fast through freight-rail system. This freight-rail proviso is met by the adoption of the Eurotunnel project. Nevertheless, I wish there had been a thorough inquiry into the scheme. What will be the actual economic effects, particularly on the regional policies?

Of course, there will be employment possibilities during the actual construction period, including the manufacture of the essential rail equipment. I also believe the opportunity for freight to save 48 hours into various parts of Europe will help manufacturers in Scotland, the north, Midlands and Wales. Of course, if our manufacturers do not rise sufficiently to the possibilities, there is also the possibility of imports from the continent coming into the UK 48 hours faster. An inquiry would have given us more positive information on the economic effects.

There will undoubtedly be effect on ro-ro ports in the south, but will there be effect also on east coast ports and in consequence on industrial and commercial development in the areas of those ports. An inquiry might have given us more firm information. Also, it is argued that with the advent of the Channel Tunnel, trans-Atlantic container ships could save four and a half days by docking at Liverpool or Clyde, instead of proceeding on to Rotterdam, thus enabling containers to be placed on fast freight trains through the Tunnel into Europe. It would have been helpful to have had sound facts from an inquiry.

Should the Channel Tunnel Bill receive Royal Assent before a General Election, it will be the Labour Party's aim to ensure that maximum economic benefits are obtained, with minimal damage to the environment. Labour also proposes a Channel tunnel Office of Fair Trading because it must not be overlooked that the Tunnel is a private commercial venture, the intention of which is, naturally, to make profits.

Commercial interests must not come before safety provisions. Indications are that Eurotunnel will be giving a great deal of attention to safety matters. However, the Commons' Select Committee, which looked into the many petitions against aspects of the Bill, listed a number of matters which the Committee said were questions of public policy and not competent for the Committee to make recommendations.

The Fire Brigades Union has raised important issues of safety from fire, and the National Union of Firemen, which welcomes the scheme for fast rail freight trains through the Tunnel, has made clear it will not relax its usual insistence on full safety standards. One of the key questions is the present proposal for passengers to stay in their vehicles on the shuttle-trains.

It appears that final decisions on safety will be made by the Intergovernmental Commission which will be advised by the Safety Authority. The Commons' Select Committee was assured by Major C.F. Rose, Chief Inspecting Officer of Railways, who is the designated British co-chairman of the Safety Authority, that the non-segregation of passengers and vehicles would be disallowed if the Safety Authority were not convinced that practical solutions had been found.

When the Bill, as amended by the Commons' Select Committee, came before the Commons' Standing Committee on the Bill, an amendment for desegregation of passengers and vehicles was lost by only a single vote, eight to nine. The Bill has now had its Second Reading in the House of Lords and is with the Lords' Select Committee which is now hearing petitioners. It will then return for the remaining stages in the Lords.

The Lords' Select Committee has made it clear it will wish to give careful attention to the issues of rabies and safety. Expert opinion on the issue of rabies has been given already by the former Head of Rabies Section of the State of Veterinary Service of the Ministry of Agriculture, Fisheries and Food, who had been consulted by Eurotunnel on measures and precautions necessary to prevent rabies being introduced.

The Kent County Council, whose fire services would no doubt be called upon in an emergency, has raised various points; will there be adequate men and equipment available if needed — if Eurotunnel staff are to handle an emergency in the first instance what training will be given — who will meet any additional costs that the county council may incur? From the reports issued of each day's sitting of the Select Committee, it is clear that the question of safety is receiving a great deal of attention, with the issue of non-segregation of vehicles and passengers being raised.

We shall await the Select Committee's Report and any recommendations with great interest before considering the Bill during the next stage.

Finance — Digging up the Money

Tom Rowland

The Channel Tunnel has been in financial trouble of one kind or another since the announcement, by the British and French governments at the end of January 1986, that the Eurotunnel consortium was to be given the go-ahead to build a link. Problems have been centred on the British side of the 37 kilometre wide strait, as was the case often before when the project got close to becoming reality.

This time it has been the insistence that the £4.7 billion estimated to be needed to get a service into operation, which must all be raised from the capital markets and without aid from either State, that has been at the root of the problems. The solid attitude of the French, who have always been more united in their support of the project, has eased the journey through the decision-making process in Paris. Further, the French appeared to have their part of the finance package lined up as early as January 1987.

The crunch comes in autumn 1987 when Eurotunnel, the bruised, battered, but still credible consortium, has to raise a further £750 million from a public share offer. If it fails to meet the target, its bid to build the Tunnel which so many before have battled in vain to complete, will be in severe difficulty.

Eurotunnel's plan is to raise a total of £6 billion. Of this, £4 billion will be in the form of conventional bank loans, a further £1 billion will be stand-by credit from the banks offering loans, and £1 billion will come from the sale of shares in the company to anyone willing to purchase them.

Until early spring 1987 Eurotunnel's London operation was behaving as if it was the project's worst enemy. Its actions could almost have been calculated to shake City confidence and scare off investors at a crucial time. Boardroom rows, top level resignations, organisational muddle and bad mishandling of some of the early stages of equity raising all combined to give an image of bumbling half-heartedness. Eurotunnel tried hard to brush up its image and its control of the project in the run up to the public share offer; given the importance of imparting an impression of calm competence to would-be investors, how successful it has been will determine the future of the whole operation.

The catch in the financial structure from the point of view of the consortium is that it cannot touch loan money until most of the £1 billion has been raised from the sale of shares. This is the money at greatest risk. The plan is to spend it before loans come on stream and the returns investors can expect are heavily dependent on everything going to plan. The £1 billion was divided into three distinct tranches in the original Eurotunnel plan. These were called Equity 1, Equity 2 and Equity 3.

Equity 1 was raised early on from members of the consortium. During 1986 they came up with £46 million. Equity 2 finally raised a further £206 million at the end of October 1986 through a private international offering — banks and cash-rich organisations around the world were invited to buy a piece of the action at advantageous rates.

It was here that Eurotunnel found itself in the most horrific public mess. It had already been badly embarrassed by being forced to postpone the issue at the last minute, and when it did go ahead, the target of £206 million was only reached once the Bank of England and the British Government had stepped in to browbeat and herd reluctant institutions into the fold.

The larger public share offering, Equity 3, was originally planned for the summer of 1987 but was later rescheduled for the autumn of the same year, this time at a more relaxed pace. A smaller offering of around £75 million to the institutions was slotted in as a curtain raiser to the big autumn push. A success here was calculated to wipe out the memory of the earlier debacle and create the right atmosphere in which to raise the rest of the share finance.

Eurotunnel also had a restructured management and a new UK chairman, Mr Alastair Morton, who came with strong credentials as the Chief Executive of the Guinness Peat Group. With his arrival the Tunnel company started to make more effective effort at getting the positive side of its project across to a still rather bewildered investing public.

If all goes to plan, Eurotunnel says it can deliver a potential capital gain by 1994 of six or seven times original investment stakes to those who buy them at their flotation, and hold on. It is confident it can persuade even the most sceptical that everything will go well. Its shares will be high capital growth investment translating into a safe utility stock once the Tunnel is open for business. Those getting in early will have the advantage of high capital growth. Later, the investment will be very safe, but one will have to pay more for the privilege of investing.

Eurotunnel has been lined up as the latest instalment of popular capitalism. Both governments hope Equity 3 will attract large numbers of small investors who will contribute a substantial part of the £750 million target. If this does happen it will be a remarkable achievement. The Tunnel is a very different kind of investment to any of the privatisations which have boosted share ownership so significantly in the past three or four years in the UK, such as British Telecom, the Trustee Savings Bank (TSB), British Gas or British Airways. It would have been impossible to even contemplate raising significant sums from small investors if the environment created by these privatisations did not already exist.

Sir Nicholas Henderson, the Tunnel consortium's first chairman and the former British Ambassador to Paris and Washington, outlined the rationale. A large equity base had been built into Eurotunnel's financial plan partly because it would encourage the banks to undertake long term lending. "Furthermore, previous ideas about the limits of the market had been radically altered in late 1984 by the success of the British Telecom flotation: besides, the share market had become increasingly bullish", he wrote in his book about the period, *Channels and Tunnels*.

In fact, between 1980 and the start of 1987, the major denationalisations more than tripled the number of shareholders in the UK, from 1.5 million to over 5 million and the proportion of shares held by individuals started to climb towards the 60 per cent figure for the US, from the UK 1981 low point of 28 per cent. According to one Stock Exchange survey, the average British individual share investment in 1987 was just under £4,000.

How far the new investors are prepared to voyage into the stockmarket remains to be seen. Two further privatisations were scheduled for 1987, Rolls-Royce and the British Airports Authority. It is possible that their appetite will be fully satisfied with these. No private flotation could possibly hope to compete with the incentives to buy that the UK Government has consistently built into each of its new issues. The principal difference between the very widely-marketed privatisations and the Tunnel is that the former were, and are, large, long established operations, whereas the latter is at best a lot of potential, all of which is yet to be realised.

The privatisation candidates had substantial asset bases which could be offered to small investors at a discount and track records in well known and very public markets. British Telecom and British Gas had the added advantage of a complete and effectively unbroken monopoly in the provision of an essential service. Certainly there was no outside threat in the areas where almost all of their turnover and profit were generated.

It was true that with Telecom, the prospect of competition in the provision of telephone services from freshly-established Mercury Communications was on the horizon at the time of the share sale to the public, but it was well understood before the sale took place that this competition would have to be built from scratch. Further, the competition could only survive if protected and nurtured by a government-appointed watchdog and would not in any case be an effective threat for many years. Liberalisation of the telecommunications' equipment market opened one of Telecom's most lucrative sidelines to real competition, but as the biggest operator, the freshly privatised giant was well placed to dominate, as indeed it still does.

British Airways was regarded as the most risky of the Government's big sell offs. It had to survive in a volatile international environment characterised by frighteningly high levels of worldwide overcapacity and a web of international agreements which severely restricted its action. On the plus side, it also had a well established worldwide business freshly tuned up and made profitable by the Government-appointed management under Lord King, and a thick portfolio of highly valuable assets with a known and independently verifiable value.

The target audience for British Airways was very different to the one identified for Telecom and Gas. They were supposed to be comparatively well off, already have substantial numbers of other equity investments and be able to finance an altogether riskier investment. In fact exactly the same people bought it as had gone for all its predecessors. This was in spite of the City gloom which surrounded the Airways sale right the way up to the date on which the share price was announced. Lord King was widely reported to have persuaded the Government to drop the share price at the last minute from £1.35 to £1.30 each. A few pence off the price of each share was hardly going to make a world of difference. The prospects of the international airline business were not altered and neither was the ability of British Airways to compete in it; but City sentiment inevitably plays an important part in all flotations. Dropping the price a bit helped to get a bandwagon going.

Investors who had already seen their Telecom, TSB and Gas shares jump in value the moment trading started on the market gauged that the same would be true of the latest offering. Experience also told them that if the shares opened at a premium, on a steadily rising Stock Exchange, the

Table 1 How the privatisations have done

Corporation	Issue Date	Issue Price	High Price	Price on Date Equity Three was postponed (7-4-87)	Yield as at 7-4-87 (%)	Price/ earnings 7-4-87
Amersham International	Feb 1982	142p	645p	535p	1.8	26.7
Associated British Ports	Feb 1983	112p	508p	470p	1.5	24.5
— First Tranche	2-83	112p				
— Second Tranche	4-87	270p				
British Aerospace	Feb 1981	150p	688p	656		
				3.7	15.	
— First Tranche	2-81	150p				
— Second Tranche	5-87	375p				
British Airways (part paid)	Feb 1987	65p	135p	128p	4.5	10.7
British Gas (part paid)	Nov 1986	50p	97¹/₂p	90p	5.4	12.6
British Petroleum	Oct 1979	363p	937p	920p	5.4	13.3
— First Tranche	10-79	363p				
— Second Tranche	9-83	435p				
British Telecom	Nov 1984	130p	263p	252p	4.2	13.5
Britoil	Nov 1982	215p	259p	258p	4.4	34.3
— First Tranche	11-82	215p				
— Second Tranche	8-85	185p				
Cable & Wireless	Oct 1981	168p	394p	371p	1.8	19.6
— First Tranche	10-81	168p				
— Second Tranche	12-83	275p				
— Third Tranche	12-85	587p				
Enterprise Oil	June 1984	185p	279p	263p	4.6	—
Jaguar	July 1984	165p	632p	575	2.3	—

chances were, barring major disaster, that the post-flotation price would continue to rise for some time. At its close, the British Airways share offer was shown to be many times over-subscribed. Blocks of shares which had provisionally been placed with international financial institutions were recalled to the pool from which private, UK applicants would be allocated a small portion of their offers to buy.

Activation of the recall mechanism meant there was sure to be a healthy premium on the part paid price when trading opened. A steady price climb in line with activity on the rest of the market in the few weeks following the start of trading ensured that the whole exercise was judged to be an out and out success. As with other sales of public assets, the British Airways flotation was quickly followed by high levels of trading in the new stock as institutions bought substantial quantities from all too willing private sellers.

It would be wrong to underestimate the level of skill required by the underwriters on each issue to ensure that everything appeared to go smoothly. Crucial among these skills was the ability to persuade high numbers of the institutional investors that the new issue was one in which they could not afford to allow their portfolios to be light. The Eurotunnel financial strategists make no secret of their fervent hope that they will, at least in part, be able to repeat the pattern of the Airways flotation, and quickly gain momentum from an apparently standing start.

But despite the jarring popularism of the "Sid" commercials put out for Gas, and the more carefully targeted appeal of the Airways publicity, one of the significant points of all the flotations has been that the same people have been attracted to them all. There have been no research findings to point to essentially different classes of investor being attracted to some of the issues and not to others. It has yet to be shown that these people are likely to go in large numbers for the Tunnel flotation. Strategists at Eurotunnel were well aware of the problems they would create in trying to persuade investors that the Tunnel was another TSB or British Gas.

The company commissioned extensive market research of its own to try to identify retail audiences most likely to be sympathetic to Equity 3 shares. Mr Ian Callaghan, treasury manager at Eurotunnel and the man in charge of much of its financial presentation, said that "there was not a very encouraging response from the young affluent, new shareholders who had between £2000 and £3000 permanently available for investments in the privatisations".Neither were sophisticated private investors who had held portfolios of equity stock for a long time over keen, according to the findings.

But there was an intermediate group of relatively well off, older private investors with a share portfolio worth between £2000 and £5000 composed of comparatively few stocks. These were people with a few shares who would be prepared to hold on for a period to realise a return. This group is known in investment analyst's jargon as "Belgian dentists" according to Callaghan. He was not sure of the origins of the name, but calculated that there are enough of them sprinkled around booming

Alastair Morton, UK joint chairman of Eurotunnel. He joined the board early in 1987 and
restored the City's faith in the leadership.

southern Britain and farther afield to provide Eurotunnel with a healthy
level of private participation.

To make sure, the advertising budget was raised above the £10 million
figure at which it was widely believed to have been set in December 1986.
The chances are that a high proportion of the money will be spent on
promotion in the financial press. The advertising agencies were commissioned to find out if "Belgian dentists" watch *Dallas*, the popular
television show imported from the USA, so that advertising through that
medium could be adjusted accordingly. If Equity 3 had taken place in July
1987, Callaghan stated that the company would not have been given any
UK television time because the Parliamentary procedure would not have
been complete at the advertising planning stage. Callaghan said "you
need at least a month to mount an effective 'Sid' type campaign".

The House of Lords Select Committee on the Channel Tunnel Bill
finished hearing petitions on Thursday 30 April 1987. Lord Ampthill
wound up the wearisome proceedings and said "the Committee will be
reporting to the House in due course. There will also be issued a special
report...". This stage was the furthest that any Channel Tunnel attempt
had ever progressed along the British legislative course; in France the
previous week, the Tunnel was given unanimous approval in the Élysée.
Slow but sure, the delays had made planning very difficult for Eurotunnel
and Transmanche Link, but then provided unexpected advantages.

The opportunity was there for television commercials to be run throughout the autumn. With more than £10 milllion to spend, the company could not fail to become highly conspicuous in the run up to the flotation. Target advertising in the major national UK newspapers started following the Easter weekend in April 1987. In France an almost romantic video was launched, emphasising the benefits of unity in Europe. All the same, Callaghan stated that those expecting to see another "Sid" would be disappointed. "I do not think anybody here has ever seen Eurotunnel as a natural public share. It is basically an institutional stock with some element of retail interest".

To maximise this retail interest beyond the Low Country drilling classes, a variety of promotional ideas were still actively considered in the run up to Equity 3. A presentation chest containing a £100 share and a certificate naming the owner as a founder shareholder was one idea. It might appeal as a gift to grandchildren. "We will need to go back and study the market in September before deciding the exact details of the kind of offers we will make", said Callaghan.

Eurotunnel is also considering putting in some sort of incentive into the retail issue. It cannot be on the level of the UK Government privatisation offers, and there is a legal problem. A way must first be found for the company to legally distribute profits it has not made. "We have not got any money to give away on the street corner", said Callaghan.

But Eurotunnel is certain to recall all of its shares, and offer ones of a lower denomination. A ten for one exchange seems most likely. There are a number of technical mechanisms for doing this, but with them all, the intention is to provide the market with something more digestible than the £34 unit with which, on calculations in May 1987, it would be faced during Equity 3.

Early days

The relative financial strength of the Eurotunnel consortium was apparent from the earliest days of competition for the contract to build a link between Britain and France. One of the main attractions for the UK Government of the Channel Tunnel Group (the British half of Eurotunnel) was the extent to which it seemed to have already mobilised the financial Establishment behind it.

Five contractors, Balfour Beatty, Costain, Tarmac, Taylor Woodrow, and Wimpey formed the core of the consortium, but they had financial cover from the National Westminster Bank and the Midlands Bank right from the start. There were also three merchant banks as advisors, although they were not part of the main consortium. Kleinwort Benson, the most respected of the privatisation wizards, switched between Eurotunnel and EuroRoute, almost overnight.

Bids had to be in by 31 October 31 1985. At the end of June 1985, while the rival EuroRoute was just getting round to announcing the appointment of its financial advisers, the Channel Tunnel Group (CTG) was busy in detailed public negotiations to sign up three major French banks and five leading construction companies as shareholders. This French contingent became a new joint company. With the Banque National de Paris, Credit Lyonnais and Banque Indosuez on board, plus five large construction interests, namely Bouygues, Spie Batignolles, Dumez, SAE Bovie and SGE, the financial credibility of the CTG group looked as if it was assured on both sides of the Channel. Having the former British Ambassador in Paris, Sir Nicholas Henderson, as chairman, helped the consortium to find French partners.

Fixing up a deal with the French was undoubtedly one of Henderson's main contributions. In the early months he spent much time talking with bankers and builders in Paris. He later maintained that securing the cooperation of construction tycoon, Monsieur Francis Bouygues, was the turning point in successfully extending the consortium across to France. Bouygues is much more than a construction company, with part ownership of a large number of other concerns. Its stake in Spie Batignolles (one of the other construction companies that joined Eurotunnel) could make one wonder whether it is in effect a contracting monopoly. It also has a considerable share in another major French construction group, Screg.

For CTG, setting up the joint company had the added bonus of focusing attention on the group's proposal at a critical time. As chairman, Henderson had the perfect platform from which to explain to an interested press in considerable detail, how it was proposed to build three tunnels and the kind of train services that would be operated through them. The link between his group in the UK and France-Manche, as the eight French partners had decided to call their collaboration, already looked like formidable opposition for their rivals to beat.

It was still opposition though. The bold and spectacular EuroRoute, with its combination of elevated motorways linked by two artificial islands on the edge of Channel shipping lanes to a short tunnel, was both a grander project and provided a drive-through link. Some form of drive-through scheme seemed to be the preference of almost everybody involved with the competition who was not directly linked to the Channel Tunnel Group. Prime Minister Margaret Thatcher was claimed to have a strong preference for a drive-through scheme. If it was to win, Eurotunnel had much ground to make up.

Bringing in the French had the advantage of creating the impression that the group was not dominated by UK construction companies. The reality was that UK board meetings were very much their party. "I cannot say that our board meetings in the early days were anything but chaotic; and I am not sure that this was simply because I had little experience of running an ill-assorted gathering of this kind. The directors did not seem to believe that their function was to proffer constructive ideas for pushing the project forward. Their tendency was to carp", said Henderson.

At the start the contractors were at odds among themselves. "There would often be shouting between them across the table", Henderson commented. "When they decided to get together before our meetings and decide upon a common line, the effect was to worry the banks and we increasingly found the two sides, the contractors and the banks, at odds".

Priced at under £5 billion, CTG's proposal had the immediate advantage of being considerably cheaper than EuroRoute which would have cost over £6.5 billion. Cheapness and the prospect of trouble-free equity raising that could dislodge EuroRoute formed CTG's strategy. By the late summer it was becoming at least a realistic possibility. "It is difficult to exaggerate the importance of our financing package in determining the outcome. I do not say that the British Government's financial advisers necessarily thought it brilliant, but it shone in comparison with that of the others", said Henderson.

As the deadline for bids approached, the Channel Tunnel Group frantically rushed around the City, the financial hub of London. A key element in its plan was to secure commitments from banks to lend up to £5 billion to fund the construction, even before submitting a bid. About 130 City institutions studied a 34 page confidential prospectus in the days before the close.

The *Guardian* newspaper reported that the group had won considerable support for its three-part financing plan on the basis that the Tunnel would win 66 per cent of all cross-Channel passenger traffic, 52 per cent of coach traffic and half of all container and freight shipments. The revenue forecast in the document was reported to conclude that the link would be generating £474 million in revenue by 1993. Tariffs, at late 1980s prices, would be £19.90 for each car passenger, £5.70 for each coach passenger and £8 for a through rail fare. Freight costs would be between £9.20 and £10 a tonne. The consortium was also supposed to be expecting to make about £2 profit from each person from duty-free sales.

On this basis the projected rewards were seen as being surprisingly good. The triple-bore Tunnel would cost around £5 billion to build, but it would be worth £20 billion when complete, assuming the most pessimistic forecasts for economic growth and projected traffic levels. This would be the equivalent of an annual return on equity of 18.8 per cent to 23.5 per cent for investors who had committed themselves well before the Tunnel was due to open.

None of this information was easy to verify, but served very well to create the impression that the group knew what it was doing and had a properly thought out and costed scheme. The image was some way from Henderson's description of a room full of construction bosses bellowing at each other. By the time bids went in, the CTG had lined up an impressive collection of financial backers in addition to those from the UK and France. These included Salomon Brothers from the USA, Nomura Securities of Japan and the Banco Commerciale Italiana.

The bids had to be in by the last day of October 1985, and the terms of the competition were underlined by the British Transport Secretary, Nicholas Ridley. "The most important requirement is that whatever link is chosen it must be capable of being financed without any support from government funds or government guarantees against commercial or technical risks", he said.

The Tunnel backers did not have the most imaginative scheme by a long chalk. There were none of the soaring bridge spans of the overhead schemes or the artificial island complexes proposed by EuroRoute. The CTG was banking on the two governments being as good as Ridley's word. With a bit of luck they would decide to minimise financial risks, which were in, any case, enormous. A tunnel cluttered up with train rolling stock might yet win the day. A decision on which of the schemes would be given the go-ahead would be announced by mid January 1986, said Ridley.

All of the schemes in the race were knocked off balance by a last minute entrant in the form of Channel Expressway backed by Mr James Sherwood, the ferry owner. His scheme was for two very large diameter tunnels to carry both road and rail vehicles. When it was pointed out to him that this was quite impracticable, the scheme was modified, after the deadline, to two large tunnels for cars and two smaller ones for trains.

In either form, the Sherwood application was based on incorporation of Japanese-designed ventilation equipment, something none of the others had, or thought would work, in the context of the Channel. For the UK Government, the big attraction of the Expressway was that it was spectacularly cheap on the balance sheet, if not in reality, in comparison to all of the other schemes.

Sir Nicholas Henderson of CTG and Sir Nigel Broackes, chairman of EuroRoute, could complain as loudly as they liked about the technical failings of the Expressway scheme and the unfairness of Sherwood being allowed to change it after the closing date, but the unfortunate fact was that the UK Secretary of Transport seemed entranced with it. Just a week after his speech accepting the proposals, a shortlist appeared. The

Duty-free sales form a vital part of the ferry companies' income. If the EEC bans such sales altogether, the balance between ferries and the Tunnel would not be altered. If Eurotunnel alone could not offer duty-free, revenue and dividends would be depleted.

Expressway had made it through. Even if, as others suspected, it was a decoy, it was still set to cause rivals no end of trouble. The scheme's promoter, Sherwood, was a prominent member of Flexilink, the ferry and harbour group campaigning to get the whole project shelved. However, his actions turned many of the ferry operators against him.

At least Eurotunnel had made it through to the shortlist. EuroRoute was there as well, still a favourite with many in both governments, but with growing uncertainty about the practicability of raising all the cash the scheme would require. The fourth contender, Eurobridge, looked very much an "also ran".

Many thought the two governments would eventually be attracted by the relative simplicity of Eurotunnel's financial arrangements. Cost overrun and time delay were perceived as being less harmful than on the more expensive EuroRoute scheme. Inclusion of UK and French national railways from the start gave the Tunnel an extra financial advantage. The speed at which investors would be repaid was higher because all the rail passengers would automatically be directed through the Tunnel. Without a large increase in cost, EuroRoute could not bring a rail link into operation for at least 18 months after its road opened.

The Channel Expressway and Eurobridge were rejected by the House of Commons Select Committee, which did not think the proposed technologies were proven. EuroRoute was portayed as a very bright idea but in practise too expensive a luxury. French members of a Joint Assessment Committee were also reported to be in favour of the Tunnel, if they could not have EuroRoute because of British fears about costs.

According to Henderson's account, the final days before the winner was announced saw rounds of bargaining at an ever-quickening pace. The winner was decided as a result of last minute trade-offs between politicians and businessmen. Technical specifications, traffic forecasts and cost projections seemed irrelevant. Ridley was cast in the role of conductor, orchestrating possible combinations and suggesting major alterations to the schemes which would, or at least might, improve their chances.

Eurotunnel dutifully promised to study the addition of a road tunnel to its scheme sometime in the future. Monsieur Jean Auroux, the French Transport Minister, and Mr Ridley eventually agreed a deal on 17 January 1986. Three days later Prime Minister Thatcher and President Mitterrand met in Lille, France.

EuroRoute had frightened the UK Government because of the expense involved, and in the final analysis officials had to concede that the technology Expressway proposed to use was far from certain to work. The Channel Tunnel Group and its French partner, France-Manche, had won the contract.

Building up the company

Within very wide limits, Eurotunnel will ultimately be owned by whoever is prepared to pay for the shares. As the company will be as deeply in debt to the banks as it is possible to get, freedom of action of the owners will initially be very limited. One strategy for a stridently Franco-British flavour was already seen to be in force in mid 1987, with even secretarial staff and receptionists in both Eurotunnel and TML offices appearing to be an equal mix of both nationalities. Half the company will be French and half will be British. A great deal of effort has gone into ensuring that the two parts cannot be divided. The *entente cordiale* has been dusted off with a vengeance, and those responsible are not hiding the fact.

France-Manche and the Channel Tunnel Group, the French and British companies responsible for the project in the first place, became subsidiaries of Eurotunnel SA and Eurotunnel PLC respectively. The basic reason for this structure that was to ensure the banks putting up £5 billion in loans could get their money back. "This two-tier structure has been adopted to facilitate the provision of security to the lending banks who will be providing the project finance", said the report produced in summer 1986 by Scrimgeour Vickers and Fielding, Newson-Smith, the stockbrokers hired early on to handle the flotation.

The two subsidiaries formed a partnership called Eurotunnel, which is responsible for supervising the building by Transmanche-Link and then will operate the service. Money will be provided equally by France-Manche and the Channel Tunnel Group, and they will divide the profits. Shares will come in inseparable twos, one in each company. They will not only be bought in pairs, but will have to be sold in pairs, so it will not be possible to build up a larger holding in one company than in its twin. This rather complicated structure is made workable by giving both of the holding companies identical boards of directors, and has the advantage of getting round all the legal niceties of French and British company law. The only effect on shareholders is that they will get dividends from two sources on each unit, one from France and one from the UK.

The treaty between Britain and France was designed to remove or bypass all of the possible points of conflict the two very different legal systems would otherwise have, regulating a structure which touches on the territory of both. The City, the banks and the financial analysts all seemed happy with the terms of the treaty, with no one pointing to its clauses or "fine print" as a cause of possible delay or conflict.

An intergovernmental commission (IGC) was set up some time ago by the two governments and officials from Paris and London. The regulations they devised for the building and operation of the Tunnel will have the force of law in both countries. As a process, this proved more difficult in the UK than in France.

The British legislation was complex. Under the procedure for Hybrid Bills (for details see Chapter 2 on Politics), any individual or group who felt they would be directly affected by the construction work, the location or the operation of the Tunnel had the right to petition Parliament. Financial implications were considerable. There was a possibility that the estimated costs would rise because individual petitioners managed to get changes built into the Act. Risk factors were increased proportionately for those investing before the parliamentary process was complete.

In France, the pitfalls to agreeing the treaty were far fewer. The treaty became law as it stood, once ratified by the National Assembly. The French Cabinet had only to approve it, the Assembly to vote, and the process was complete. Late on the night of 22 April 1987, this was achieved. "The joys of a centralised administration.." muttered the CTG in London as their Bill was bickered about and examined in minute detail by parliamentary committees. Only after Select Committee procedure had been completed could the Channel Tunnel Bill continue on its path and the treaty be finally ratified.

The capital structure of Eurotunnel was designed to dovetail with the UK legislative timetable. The £6 billion sourced from both bank and private investment will make investors carry the bulk of the risk, but will share the dividends between them. They will also share legal ownership of the company, each share carrying an equal fraction of the voting rights.

Arranging details of the loan package with a group of 130 or so international banks seemed an interminable process to all the finance professionals hired by Eurotunnel. For investors the outcome is important. What Henderson described as commitments from the banks turned out to be nothing of the sort. They had agreed in principle to lend, but only if stringent conditions were met. The consortium found itself with the task of negotiating these conditions after Henderson stood down from the chairmanship, following the award of the contract.

There are many different shades of grey between a good deal extracted from the banks and one that is obviously awful. As negotiations inevitably took place in private and the interests of contractors, the investors, the banks and Eurotunnel itself are all different, judgement on how good a deal it is will have to be suspended until the final financial details are released just before the main share flotation in autumn 1987. Given these risks and possible hazards, at first sight it seems odd that anyone should be prepared to invest in equity at such an early stage.

Equity 1, worth £46 million of the £1 billion total, was issued to the founders of the consortium, that is the ten construction companies and the five banks who piloted the idea through to the point where it won. They were allocated 2,652,000 units. Effectively these were cheap shares. The founders maintained they had taken the greatest risks and had already laid out quite considerable expenses.

Equity 2 was very different. The aim was to place £200 million worth of shares with large institutional investors. This was planned for sometime after the Second Reading of the Channel Tunnel Bill in the House of Commons, the point being that Bills which get through the Second Reading usually reach the statute book, even if some of the details are changed later on in the parliamentary process.

There was a risk that the provisions of the Bill could be so altered that the whole project would be made unworkable. For this and other reasons the consortium had far more trouble than it could have possibly bargained for with Equity 2.

For institutions outside the founding consortium the incentive for investing in Equity 2 was the opportunity to get some shares cheaply — Not as cheap as the consortium itself at Equity 1, but still at a considerable discount on the price the company was promising it would get when it came to the far larger public offering at Equity 3. The latter is designed to cover the balance of the equity target by raising £750 million by public subscription in France, the UK, and from international participation in the project. Only after the treaty had been ratified could the sale of the third tranche go ahead. And only after the share sale had been successfully completed could the main construction work start.

The discount structure was designed so that the earlier investors committed their money to shares, the less they would pay for them. Stockbrokers retained by Eurotunnel said that Equity 3, the only remaining issue, would be at a premium of some "40 per cent to Equity 2". This is of considerable significance to all existing investors and those considering joining them.

The relationship between the price of the three main equity issues is best explained in terms of an arbitrary index. If the price of Equity 1 is indexed at 100 a unit, Equity would 2 cost 120 and Equity 3 would be on sale for 170 a unit. All units carry rights to an equal-sized portion of the dividend stream of the company.

Equity 1 and Equity 2 made their target prices, in terms of both unit prices and the amount of cash raised, even if it was after a desperate struggle. But the exact value of these holdings is dependent on the price of Equity 3 units. City analysts expected Equity 3 to be at a premium of 42 per cent to Equity 2. Many were under the impression that this was guaranteed when Equity 2 was being placed.

Doubt exists at the accuracy of this impression. The 42 per cent figure was, it appears, not legally binding, giving Eurotunnel the option, if there appeared to be problems at Equity 3, of selling below the relative figure of 170. Many shares would be available at a slightly cheaper price than those buying into Equity 2 had calculated. Their holding would be diluted and they would hold rights to less of the dividend stream than they had expected.

Flemings, Morgan Grenfell and S.G. Warburgs, the merchant banks in charge of the public flotation, know it is not the way to make financial friends. In a buoyant stock market, with potential investors already having a favourable impression of the project, the hope is that it should not be necessary to dilute holdings of the existing investors. That is the hope. Even in the most favourable conditions for a flotation, the structure of the Eurotunnel equity package meant that there would be a degree of tension and some conflict of interest between those institutions which had been persuaded to subscribe to Equity 2 and those considering taking part of Equity 3.

If the merchant banks and stockbrokers in charge of Equity 3 could be panicked into issuing more units at a lower price than was calculated at the time of Equity 2, then in the short term those who subscribed to Equity 2 would effectively be paying more of the project than they anticipated.

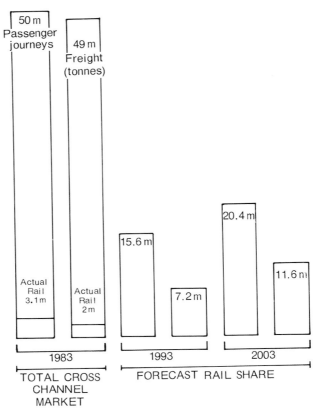

Graph of traffic growth forecasts. Source Eurotunnel.

They would also face the prospect of less of the eventual dividend stream coming their way than they thought would be the case.

Conditions have been far from perfect for the project's steps along the equity path. It has made pressure on Equity 3 underwriters proportionately greater. Structural problems of this sort are not uncommon with large flotations. In most cases they amount to little more than professional jealousies which are played out between sellers and the prospective purchasers about marginal adjustments to the price.

During the early spring of 1987, many City analysts who would be responsible for advising their clients about the value of Eurotunnel investments felt that on this occasion the results might be more serious. There was more than marginal bickering about price at stake. Some argued forcefully that there was a real possibility that the underwriters of Equity 3 would be forced to dilute heavily in order to place all of the units.

Executives at Eurotunnel were quick to dismiss such sentiments as market bluster. Remember British Airways, they said.

Private investors could get a good indication of how the institutions are reacting to Equity 3 by how closely the final share price sticks to the figure of a 42 per cent premium on Equity 2. "Fund managers are used to making investment decisions about operations which have a long track record. Even then they do not make a move without looking at all the advice available from the market and the host of analysts. In the case of Eurotunnel, this kind of background is just not there", pointed out Mr Richard Hanna, analyst at Phillips and Drew.

A large number of figures have been produced on the costs of the project, the return investors can expect and the traffic forecasts. As this is a one-off building programme, Eurotunnel inevitably has been responsible for many of the original calculations, either directly or through information it has supplied to its own banks or brokers. Some information, or at least the method of calculation, is still valid from the 1974 attempt, when all advisers were paid by both governments. Much of the City does not bear this in mind and comments merely that reliance on mainly Eurotunnel statistics increases the level of trust required. British financial institutions are still, on the whole, run by very straight, unimaginative people. If they see blue-blooded brokers working alongside heavyweight merchant banks, many would-be investors seem happier to be cajoled into subscribing to an issue of which in other circumstances they would be extremely wary.

In this respect, Eurotunnel was rather unlucky with its merchant banks. For much of the share flotation process, Morgan Grenfell was under the cloud created by the apparent insider trading within Guinness. Eurotunnel had to struggle at one stage to avoid being tainted with the fallout. Flemings, its other banker adviser, is not widely perceived as having the same track record in successful completions of flotations of this size and scale, which is not to cast doubt on its ability to do it, but merely to record that some of those to whom it must sell substantial blocks of Equity 3 voiced that worry.

Kleinwort Benson, the king of the big flotation, has been mentioned a number of times in this context. Many brokers were convinced by rumours that it had been asked to join, but had refused, although having changed camps to loser EuroRoute, it could be difficult for it to prodigal-son-it back to Eurotunnel.

Some others felt that Eurotunnel's stockbroker team could have been stronger. "If Cazenove comes around and asks you to subscribe to something like the Tunnel, you take a piece so as not to be left out next time, when the issue on offer from them is something you really want badly", said one outsider. In the event, neither of these two names was involved in mid-1987.

Eurotunnel acknowledged the fears to some extent when at the time of postponement of Equity 3, its UK joint chairman, Mr Alastair Morton, announced that a third merchant bank, S.G.Warburg, had joined the other financial advisers. The stockbroking side was also strengthened by the addition of Warburg's own broking division.

Many of those who had been vociferous in their criticisms of Eurotunnel would also concede that if the City does manage to raise all of the finance for Equity 3 without further hitches, it will do itself a lot of good.

There were an estimated £50 billion worth of large construction projects in the planning phase in different parts of the world at the start of 1987. Much of the business of raising the finance would drift to London as a direct result of the two sides of the Channel eventually being linked in a £6 billion privately financed project. It is the sort of track record that could earn healthy commissions for many more than a handful of brokers and banks through the placing of this follow-on business. Hong Kong's two major cross-harbour tunnels proved an excellent investment, with financiers there clamouring for the right to contribute to the cost of the MTR, an underground railway system as well. The Europe–Asia bridge links across the Bosporus also produced fierce competition for the right to fund them.

The problem was that Eurotunnel was not doing much to stop itself being caught in the tangles that the short term perspective of much of the City breeds. Fund managers traditionally have not been very good at looking at anything longer term than three months; they have a preference for stability and want to back projects that look strong, confident and have lots of positive public relations.

Eurotunnel was bound to be difficult to sell. Leaving aside the financial details of the project, right up to the early spring of 1987, the consortium had done a poor job on projecting itself and the scheme, in terms that those it would shortly be asking for £750 million would appreciate. Peers complained in the House of Lords that their letters had not been answered, even journalists said that their telephone calls were not returned; many minor slights added up to a bad press and resulting bad reception in the City. However, for a few days before the close of Equity 2, the marketing mix was said, by the City institutions on the receiving end, to be very potent.

The idea of Equity 2 was to raise a quick £206 million from selected international financial institutions, to provide a solid cash basis to tide the consortium over during the twelve months before planned flotation of Equity 3, then scheduled to take place in summer 1987.

Fumbling the issue

Well over a year before, in the spring of 1986, as preparations for Equity 2 got under way, it became increasingly clear that no matter how sincere the wish of the UK Government to have the private sector take care of all the financial backing, it was going to have to become involved. Assuming the whole of the public flotation goes smoothly, events so far will ensure that the finished Tunnel stands as a monument to the successful intervention of the State in the financing of a large infrastructure project.

Mrs Thatcher's insistence at the outset that the whole programme should be completely privately financed came to grief over Equity 2. The £206 million would not have been forthcoming if the Bank of England and Ministers had not set out on a programme to entice and persuade institutions into coming up with the cash. But their task would not have been so difficult if it had not been for determined no-holds-barred opposition from the ferry operators grouped together as Flexilink. Despite its actions resulting in complaints to the UK Advertising Standards Authority, they were certainly extremely effective.

As the principal condition insisted on by Mrs Thatcher over French objections was that the Tunnel be built with only private finance, the extent to which this proved possible is itself of interest.

To the deep embarrassment of all involved, the deadline had to be extended from the summer of 1986 to the autumn. Eurotunnel would doubtless prefer to forget the period, but as it is the only occasion on which the credibility of the project has so far been tried in the markets, it is worth looking at in a little detail, if for no other reason than to test Eurotunnel's contention that the events of autumn 1986 were not a good indicator of what is likely to happen at the far bigger public flotation.

The official prospectus outlining the project and Equity 2 was due to be dispatched to fund managers in early June 1986. Flemings, Morgan Grenfell, Scrimgeour Vickers and Fielding Newson-Smith had approached around 200 UK institutions before the publication. All had expressed interest, and were holding off making a decision until they had seen the document, the outside world was told.

During June, details of Equity 2 started to leak. On a Monday morning in mid-June the new Eurotunnel chairman, Lord Pennock, formally launched the campaign which would reach its height with the £206 million share placing.

From that moment on it was hard to find a single newspaper commentator, stockbroker or analyst who recommended the issue. It was open season for articles seeking all the flaws in the Tunnel. One parliamentary correspondent confesses "I am actually in favour of the Tunnel, but it is so much easier to write stories against it. Editors do not really want good news".

The risk was too high, the return too low, the market too full with other new issues, they said. Attention was focused on the impact inflation might have on the project during the construction phase. The market liked the fact that tolls had been designed to be inflation proofed, but it worried about the risk from inflation in the phase before the Tunnel was complete. It also worried that the UK construction industry did not have a good reputation for finishing on time. If the 1993 season was missed because of late completion, then the potential returns would shrink dramatically, warned Andrew Taylor of the *Financial Times*.

Eurotunnel had well-rehearsed replies to major criticisms. It would point to a string of more recent projects completed on time and within budget by the UK construction business, and it had built elaborate fail-safe mechanisms into the construction contracts it proposed to hand out, in order to minimise the risk of overruns. It could have legitimately pointed out that the risks of not finishing on time were really very small, whilst the rewards available to investors could make brokers' assumptions ludicrously pessimistic in the far more likely event of the Tunnel opening on target and then traffic flows building up at a healthy rate.

Lord Pennock, as UK chairman, was effectively figurehead and main spokesman for the company. He did not do well in getting this message across. Eurotunnel's public relations during the summer were close to abysmal. Where it should have emphasised the sheer size and ultimate money-making potential of the scheme, it took refuge behind masses of difficult-to-verify figures. Where it should have produced sharp rebuttals of specific criticism, it blustered or said nothing at all.

The behaviour of the UK Independent Broadcasting Authority (IBA) over television advertising of the Tunnel was a case in point. Early in July 1986, the IBA announced that it had changed its mind about allowing Eurotunnel to put out a 60 second commercial, timed to coincide with Equity 2. As the Bill enabling work to start was still going through Parliament, the IBA decided that the advertisement contravened its code of practice. This disallows any advertisement which relates to current public policy.

The IBA had originally given Eurotunnel the go-ahead. This change of mind was an obvious disappointment. But the consortium hardly made a virtue from adversity. It could have started newspaper and magazine advertising. Shown the "facts", the IBA would not let the public see. Instead it fielded Pennock who could only say how disappointed he was, complain that British Telecom was allowed to advertise during its

privatisation campaign, and make a rather arcane reference to the press
having more freedom than the broadcast media.

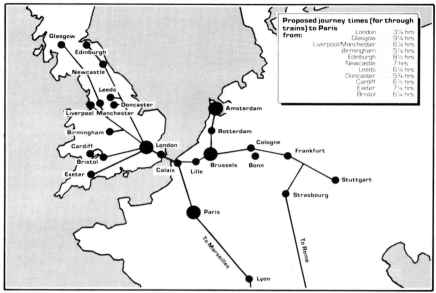

Proposed journey times (for through trains) to Paris from:	
London	3¼ hrs
Glasgow	9¼ hrs
Liverpool/Manchester	6¼ hrs
Birmingham	5¼ hrs
Edinburgh	8½ hrs
Newcastle	7 hrs
Leeds	6¼ hrs
Doncaster	5¾ hrs
Cardiff	6½ hrs
Exeter	7¼ hrs
Bristol	6¼ hrs

Map showing proposed through-train times to Paris for journeys through the Tunnel in
1993, from major UK Cities. Use of high-speed trains on UK routes will increase the
advantage over ferries and even hovercraft, still further.

Arguments about television were, in any case, a sideshow. By early July,
many were starting to wonder what had happened to the *Pathfinder
prospectus*. Its publication was now late. The stockbrokers would only say
they were still working on it and "there is no firm date".

To outsiders it was becoming increasingly clear that the behind-the-
scene talks between Eurotunnel's financial advisers and the institutions
had not gone well. On 9 July 1986, the Board of Eurotunnel called an
emergency meeting to discuss postponing Equity 2. The formal
announcement was made a couple of days later. In the meantime,
Eurotunnel was roasted by the press.

"Ostriches stick their heads in the ground to avoid seeing the obvious.
Eurotunnel lacks even a hole in the ground to hide its blushes as its
optimistic timetable comes another cropper", ran the introduction to a
story in one daily newspaper.

Fielding Newson-Smith, which had now changed its name to County
Fielding as part of the preparation for the deregulation of the City,
insisted that the delay was not particularly serious. Putting together the
prospectus had been long and complicated, and translations into several

languages were needed, it explained. As there were still a few minor matters to be sorted out before the offer document would be complete, it had been decided to put the launch back until after August 1986.

As 35 per cent of the cash had to be raised in France, and the Paris business market was virtually shut down during August, the sensible course was to wait, argued the brokers, as they did again for Equity 3. The French elements in the whole of the financial package had been meshing together like a well-oiled machine; still in any joint poject it is fairly safe strategy to blame the other side when things go wrong.

London and Paris both had to deliver £70 million. A further £20 million was to come from the US and £20 million each from the rest of Europe and Japan. County Fielding scoffed at suggestions that it was going to be difficult to raise the £70 million in London because of the number of other share issues about to hit the City. "The amount of other new issues has nothing to do with it. The sum of £70 million is small in the UK speculative equity market", said a spokesman.

Small or not, at the same time as the announcement of the postpone-ment, Eurotunnel made a series of boardroom changes aimed at strengthening its financial expertise and to better prepare the path for the *Pathfinder prospectus*. Michael Julien left his position as group financial director of the Midland Bank to become a full-time member of the Eurotunnel board. A total of six new directors were to come into the company, including Sir Nicholas Henderson, who had left only a little time before, when the company won the fixed link contract.

It also emerged that the "details" that had been holding up the *Pathfinder* prospectus were questions from the 40 banks putting up the loans. "The concession terms have been agreed and the banks are seeking clarification of some of the terms", Lord Pennock said. "It is mainly to do with the construction contract and the calculation of capital costs".

In fact the syndicate had underestimated the complexity of the negotiations with the 40 banks. They seem to have been fairly happy by this stage with the overall economics of the project, but they did not like either the corporate structure or the makeup of the board of Eurotunnel. Most of this had been sorted out, but what Lord Pennock called "contractual details" had not.

Arguments were caused by the contractors attempting to increase the amount by which they could exceed the target price before running into penalties, the agreement with the two national railway companies which was stuck in wrangles about minimum usage levels, and worries the banks had about the accuracy of Eurotunnel's traffic forecasts. The banks were keen to push more of the financial risks on to the contractors.

Any dispute was kept well concealed, but potential difficulty in raising the £270 million of Equity 2 expected from London was indicated by a poll carried out by the *Financial Times* newspaper between the 23rd and the 28th of July. Less than a quarter of the 25 institutions approached were going to seriously consider investing in the project. Ten out of 25 fund managers said they would definitely not be investing in the scheme and a further nine had yet to make up their minds. Eurotunnel was still saying that of the 40 institutions it had approached the previous autumn, 30 had agreed in principle to support the issue and later on the flotation. Even on the basis of its own far more gloomy poll, the *Financial Times* was sure the consortium would have no trouble raising the £70 million during the autumn.

As September drew near, Eurotunnel proved as good as its word. It was expecting around 30 institutions to take the £70 million of stock between them. It was even bold enough to say that 20 of its 30 were in insurance. On 26 September the 106 page *Pathfinder* prospectus finally made it into the public gaze. It was not officially due to be published for another three days, but the outline Eurotunnel gave did not get too bad a reception, although none of those who had come to oppose the project seemed likely to instantly change their minds. At least now there were some official projections that could be examined, and the assumptions tested on which they were based.

In 1994, the first year of planned operation, there would be a turnover of £752 million, pre-tax profit of £109 million and a dividend of £3.34 for each of the paired units. By the year 2002, the pre-tax profits were to rise to £727 million and the dividend to £11.82 per unit. By 2041 the profit would top £12 billion on a turnover of £15 billion and the dividend ought to be more than £200 per unit. Eurotunnel was looking for a total of £206 million, not £200 million, the document revealed. For all those considering buying Equity 2, far more than this slight change and certainly more important than the projections of profits ten, twenty or more years ahead was the initial rate of return. On the basis of all of the assumptions in its economic model, the *Pathfinder* was promising investors a return of 16.7 per cent on their money. It became obvious that the insurance companies, pension funds and other institutions on the Eurotunnel hit list had been distinctly unimpressed with the rate they could expect. It was being very badly received, and it turned out to be the most significant stumbling block during the whole of the placement. A press briefing held on the same day by Flexilink set out to criticise and ridicule the prospectus. The mood seems to have been summed up by a research paper put out by stockbrokers Phillips and Drew the day after the official launch of *Pathfinder*. Entitled "Digging Deep for Eurotunnel", it questioned whether the modest return was worth the risk. "Even if the base case assumptions are right, these returns are low by venture capital standards and indeed are below the UK industrial average in the last ten years", it says. Richard Hannah, transport analyst at Phillips and Drew, went on to dance on the spot he seemed to have marked out for the grave of Equity 2. The average UK company produced a return on capital of 18.9 per cent for 1985, he argued. During the previous ten years, at 22 per cent the average rate of return in the British equity market had been higher still.

Most of the drama of the shortfall was inevitably played out behind closed doors. The two banks and two brokers acting for Eurotunnel had until the end of October to raise their quota, £70 million. It became clear that some very big guns had been called in to make sure that the deadline was not missed. None of the parties involved has been prepared to talk openly about the exact details of what was said to them, so it is hard to be sure when the Bank of England first started to help herd the reluctant into the fold. It was certainly before the possibility started to be discussed in the press. In retrospect, the divergence between what was going on in public and developments behind the scenes can be seen to have been getting wider as the deadline approached.

As late as a week before the close County Securities was grabbing headlines by publicly saying that the target would be reached without serious problem, and the institutions had been convinced by their latest analysis, which showed the rates of return quoted in the *Pathfinder* to be very conservative. This they said was because non-Tunnel-traffic-related money-making activities had not been included in the calculations. How much of the intervention could be described as legitimate lobbying by the Bank of England and how much was straight bullying by the UK Government it is difficult to say, but direct and highly effective intervention there certainly was. Two days before the official deadline the best available guess was that the shortfall was between £10 million and £15 million, leaving aside the danger of the whole house of cards collapsing, once those who had provisionally committed themselves became aware of how exposed they could be by taking up the shares.

The Paris, New York and Tokyo placings had all gone well. Tokyo and Paris had no problems to report with the issue even generating modest enthusiasm. New York was a little more problematic; the shares had been placed well before the deadline, but there was a danger of the commitments falling apart if it became apparent that London had failed to make its quota on time. Officials from the Bank of England could legitimately point out the damage failure to come up with what was a comparatively modest sum would have on the international credibility of the City. London had built a solid reputation for its successful handling of large and potentially tricky issues over the period of the big privatisations. Mucking up Eurotunnel would do this reputation no good at all.

The credibility of the Eurotunnel financial advisers had in any case acquired an unsightly dent. On 27 October The *Daily Telegraph*, *The Times* and the *Guardian* carried stories speculating on the future of Morgan Grenfell and Scrimgeours. Both could shortly be replaced as advisers on the Tunnel, they said. Kleinwort Benson, Cazenoves and Barclays Bank were all tipped as the most likely replacements, despite at least two being involved with previous fixed link rivals. The position of Lord Pennock was also thought to be unstable. Rumours that he was about to be replaced by Sir Nigel Broackes proved to be wrong. There was, all the same, a very good basis to all the speculation. Having been called in to help, the Bank of England wanted changes to the board of Eurotunnel. Sir Nicholas Henderson has already described in his book *Channels and Tunnels* how the UK Government went to a great deal of trouble to encourage some form of amalgamation between the competing fixed link schemes. In essence, two of three, with Eurobridge treated as an outsider, should gang up on the the others and win the mandate for their trouble.

Although the Government said it would force the consortia to join up as a condition of awarding the contract, in the final scramble for the winning post a lot of pressure was applied to get them to do so of their own accord. After the ploy had failed and the Government had been seen to keep its promise of making a clear choice in favour of one of the contenders, it was again free to pursue the policy of amalgamation. To officials, who had seen greater financial stability as the main benefit of an enlarged consortium, the fiasco that Equity 2 was turning into was ample justification for having another go at trying to form one. There were also renewed rumours that Kleinwort Benson, the principal bank in the EuroRoute consortium, was negotiating a possible later re-entry to the stable of merchant banks. Eurotunnel denied that any shake-up was under conderation, but fuelled the speculation by refusing to say if it was considering bringing in any other financial advisers to help with Equity 3.

On 31 October 1986, the announcement was finally made. The Equity 2 target had been reached. A minimum of information was included which did not even state how much had been raised by the various countries. Sighs of relief were more in evidence than much celebration, but at least the money was in. When the exact figures were released they showed that the City had in fact exceeded its target and raised £75 million.

The French firmly believed the British had made life complicated by setting overtight deadlines and creating credibility gap in the process. Once a deadline was getting close, naturally enough, all the potential investors were bound to wonder what chance there was of construction being done on time even if the finance stages were in danger of being late. The number of institutions had been more restricted in France, which meant private deals were easier to set up and achieve with the minimum of trust. There was lot of irritation in Paris over the unsightly mess in

London's handling over what was, after all, supposed to have been a private placing.

Eurotunnel had to buy the 1974 tunnel workings from the UK Government before excavation could start again.

As an analysis, this is in some respects unfair. The four London issuing houses had to combat the notoriously short term investment horizons of many of the UK institutions. A large slab of UK manufacturing industry had already sunk because it had failed to get the institutions to look more than a few months ahead, so an easy passage for a 55 year infrastructure project with its first return in 1993 was a bit of a tall order. And, however unjustified some of its comments were, Flexilink, like a nagging appendix had a noticeable detrimental effect in the UK rather than France. Journalists fanned the flames. Eurotunnel's financial advisers did not have the advantages of their French counterparts, who were dealing with a smaller number of more malleable institutions, many of which were nationalised.

Tight deadlines dogged, rather than eased, marketing of Equity 2, which could not begin in earnest until the wrangle with the banks over contract details was out of the way. These negotiations carried on until the end of September. Even with the postponement of the placing, it left very little time in which to subject all the targeted institutions to much high pressure marketing. In the aftermath, when it was certain all the money was in, the consensus among the institutions was that Eurotunnel had done a reasonable job of selling the project to them, given the shortage of time and the impact the anti-campaign from Flexilink had been having. During the summer and autumn of 1986, few people in London could have failed to notice the large posters and full page newspaper advertisements put out to counter the concept of a cross-Channel Tunnel by the grouping of ferry companies and port authorities.

If the effect of the Flexilink campaign was damaging in terms of broad public relations for Eurotunnel, in the City the impact was magnified. At each stage of the placing, bright, well-put-together Flexilink propaganda arrived on the desks of fund managers soon after any initiative made by the Tunnel's financial advisers. Although it probably did not on its own change any minds about the project, the anti-Tunnel material added to the climate of uncertainty, and opposing statistics or forecasts were difficult to dispute without appearing paranoid or developing a public slanging match. The climate was not improved by the post-mortem into Equity 2. When the full list of subscribers was eventually published in 1986, it showed that the consortium would have been £12 million short of its target if the founding banking and construction shareholders had not come forward to increase their stake.

It had not been part of the plan for the subscribers to Equity 1 to take part in Equity 2 as well, although it was not against the rules. Their participation in the second stage also explained the mystery of why £75 million had been raised in London when the target was only £70 million. The Equity 1 shareholders had in fact taken £20 million of Equity 2. The largest UK shareholders who were not part of the Equity 1 group were insurance groups Standard Life, Legal & General and the Prudential. They had all subscribed for between 2.3 and 2.9 per cent of the 8.58 million shares. In France, the largest holding had been taken by Compagnie Financiere de Suez, which took 4.5 per cent of the total shares on offer. Eurotunnel said the increased size of the London subscription had been necessary because sterling had weakened against the franc just before the placing. An adjustment was required if the relative sizes of the UK and French holdings were to be kept in balance. Provision had been built into the original financial structure for the founders to step in and take up the slack, if currency drift created a shortfall in the few days before the close of the placing. It proved to be a convenient mechanism to use to take up all of the extra left by the shortfall in institutional demand.

Mr Colin Stannard, a senior merchant banker on secondment to the project from the National Westminster Bank, blamed the pension funds and their financial advisers rather than the insurance companies for the bulk of the problems that beset Equity 2. The lack of institutional investors had certainly taken the consortium aback, he admitted in an interview given a few weeks after the target had finally been reached. It was "astounding" he said, that the merchant banks advising the pension funds to stay away were not prepared to back an unquoted investment, despite the fact that a full stock market quotation was planned in less than a year. "It was a shock to us that they were not prepared to wait nine months, when the banks [funding the project] were ready to look forward 18 to 20 years".

Stannard also confirmed how nearly the whole placement was jeopardised by the lack of interest in London. Japanese and particularly US investors were perfectly happy to buy the stock on its potential for capital growth, but were getting cold feet in significant numbers when the home investors appeared reluctant, he explained. It is arguable that the banks responsible for the loans were in a very different position to the fund managers being asked to put up equity. For a start they were not facing the prospect of seeing their investment go down to a worthless curiosity if anything went wrong. But if it was the case that pension funds were put off because there was no immediate market in the shares, Eurotunnel could legitimately claim that with Equity 2 out of the way the worst was over. After Equity 3 was raised, there would be no waiting before the shares could be traded.

Borrowing £4 billion from the banks and negotiating a further £1 billion in standby finance required complex negotiation. Eurotunnel's finance has already been held up once because of protracted arguments between the contractors and the banks over details of the package, which lays down at precisely what stage each of the elements in the loan package can be spent, plus the financial responsibilities of the contractors if the project is delayed and of the banks if the costs start to overrun. Much of the consortium's time in the period between the completion of Equity 2 and the start of the campaign to market Equity 3 was taken up in further negotiating the detailed terms of the loans with the banks. Lending banks will have to sign a legally-binding agreement before Equitty 3 can go ahead. One of the risks faced by subscribers to Equity 2 was that the bank loans could fall through.

A term sheet outlined the conditions on which the loans would be made available.

1. The Treaty and Tunnel concession have come into force.
2. The construction contract and main agreement with the rail companies on both sides of the Channel are settled to the satisfaction of the banks.
3. At least £650 miliion of the equity has been raised.
4. At least £600 million of the equity will already have been spent on the project.

It meant that the loans would only become available once Equity 3 has been successfully completed and over a third of the money raised through it has been spent. The loans would then be handed down in blocks as the building work needed new funds. Money will be paid back to the banks by regular deposits, as detailed in the schedule. The first repayment must be made three months after the Tunnel starts to take money. On this basis the loan should be paid back to the banks by mid 2002. The term sheet lays out a quarterly schedule of repayments. But if the Tunnel is doing well it will be able to pay back part of the money at a faster rate.

This would be very much to the advantage of the equity investors, as the more quickly the loans are repaid, the lower the interest charges that will be due, and the more residual funds left for distribution to them. Quarterly repayment will be the greater of minimum repayment figures for the relevant quarter, as defined on the schedule, or a fixed percentage of the net cash flow of the project. Eurotunnel's calculations assume a turnover rising at a rate which allows for a significantly faster payback than the minimum laid down in the quarterly repayment schedule. In the unlikely event of Eurotunnel's takings and turnover being so far below the expected level that it defaults on the repayments the banks will have the right to take over all of the Tunnel assets and appoint a new operator in place of Eurotunnel. As they hardly need to be told, this would be bad luck for the equity owners of Eurotunnel; and calculating the size of the risk of it happening has to be part of the homework of all those considering investing.

The scheme could still go ahead if all of the planned £750 million is not raised. As long as £400 million is found to go with the £250 million of equity already in the company, the threshold outlined by the banks as a condition of the loans being drawn down will have been reached. This at least was the position at the time of Equity 2. It may well be that one of the changes which the banks insist be included in Equity 3 is that the full £750 million be raised before the loans can be touched. Under Equity 2, any shares left over if the £750 million target was not quite reached would be taken up by the banks in the form of warrants. The number of warrants issued would be based on a formula related to the size of the shortfall. In effect, the bigger the equity stake held by an outsider, the smaller the equity involvement of the banks would be. All of the warrant clauses may have been removed by the time the Equity 3 prospectus is unveiled.

One thing is certain: all of the loans will have to be repaid. Even on an accelerated rate, repayments would get in the way of healthy-looking dividends for the equity holders in the first years of operation when interest on the loans is at its highest, so provision has been made in the loans agreement for Eurotunnel to refinance all of its debts after paying-customers start to arrive. It would do this by issuing bonds on the international market and using the revenue to repay the bank loans. The tunnel company will be free to issue bonds from two years after the completion of the building work. Assuming work finishes on time, this means the first bonds could be issued by early 1995. At the time of Equity 2, Eurotunnel's financial advisers said that tapping the bond market would make it possible to repay all of the loans by mid 1999. The consortium estimated that it ought to be just under £4 billion in debt to the banks by the time the first shuttle is ready to run through the completed tunnel.

At January 1986 prices it estimated the total cost of the project as £4848 million. The £1 billion equity will be spent first, leaving a total loan facility of £5 billion of which only £3.85 billion will actually be required, if all goes to plan.

At the time of Equity 2 the estimated cost schedule was as shown in Table 2.

Table 2

Project costs	£m
Construction and equipment	2523
Owners' costs	296
Optional investments	101
Adjustment for inflation	660
Interest capitalised during building	1152
Finance fees and costs	114
Total Costs	4848

The costings were based on an assumption that inflation would be 3.5 per cent in 1986, 5 per cent in 1987, 5.5 per cent in 1988 and 6 per cent for the rest of the building period. Although the contractors are subject to a penalty of costs overrun and bonus arrangement give them an inducement to undershoot an admittedly unlikely overrun in excess of 20 per cent falls squarely on the company. If construction problems drive costs above that point, the equity would be squeezed and the debt would pile up. To make things worse overruns could be associated with a delay to the start of traffic flow.

Just before the start of Equity 2 the *Financial Times* gloomily pointed out how even a relatively mild case of overspending would make a painful difference to the attractiveness of the investment. On Eurotunnel's own projections, a one year delay and a £200 million overspend (only 4 per cent above budget) would put back the first dividend payment for four years, lift peak debt by £750 million, frustrate the planned securitisation of bank debt and drop the overall rate of return to shareholders by a tenth. So, the ability of the consortium to keep the cost firmly in line with the plan is vital to the equity investors. The Tunnel financial package is indeed delicately balanced, and one of the early casualties of even the

John Reeve, Director General of the Translink joint venture of contractors. When the Tunnel is built, all the contractors will have been well paid for about eight years of work, will continue to receive dividends from their shares in Eurotunnel and will doubtless thoroughly exploit all the other house building and infrastructure projects that will stem from the scheme. Astute share buyers will probably buy into the individual companies as well as into Eurotunnel.

slightest increase in construction costs will be dividend levels. Eurotunnel's philosophy is to try to keep costs in line by giving the contractors heavy financial incentives to police themselves and come in on budget. The contractors stand to make really spectacular profits if they deliver at the agreed price or below it, more than they could expect to get from normal tendered work. But the penalty payments will not make up the losses that equity investors stand to see, if things go badly over price. Investors would have the compensation of knowing that the construction companies had been well and truly beaten about the financial ring for their failure, but the company would have to borrow extra cash to fund the extra bills, and deduct the increased interest charges from the dividend pool.

Safeguards to encourage the contractors to keep to the deal are elaborate, but investors still have to decide if they think they will work. The aim is to keep the building costs in line with estimates, by giving contractors lump sum payments for specific parts of the work which they have to complete for the agreed price. Of the overall bill, 40 per cent of the payment should be on this basis. For an even bigger portion, 50 per cent of the total contract, a very different system is proposed. The contractors will be given in advance cost targets and will have to pay stiff penalties if they go over budget. It is here that the biggest headaches could occur — more of which later. The remaining 10 per cent will be subcontracted after competitive tender.

The main contractors are the ten construction companies in the consortium, responsible for designing and building the Tunnel complex, which will then be handed over to Eurotunnel to operate for the period of the concession it has been given by the UK and French governments. The stockbrokers responsible for selling Equity 2 to the City institutions pointed out that most of the notoriously late and over-budget UK construction projects in the past have had a split between design teams and construction companies. With the Tunnel they will be one and the same, and the hope is that as a result many of the demarcation disputes which troubled the UK industry in the 1960s and early 1970s can be avoided.

Having the whole package designed and built by one integrated organisation has obvious advantages, but by the same token there are pitfalls. The construction companies doing the work are also part of the consortium. Eurotunnel does not try to disguise the conflict of interest. The sales pitch of the financial advisers does its best to make it into a positive virtue.

If they delivered a working tunnel on time, the contractors would earn profits "which are attractive compared with those currently being reported from competitively tendered work", said Scrimgeour Vickers and Fielding, Newson-Smith in a masterful understatement included in their Equity 2 analysis. But the equity investors would have been delivered a bright new tunnel from which they should make attractive returns, the brokers went on to argue.

"If the contractors are late or over budget in delivering the package or if it fails to perform within the specifications for shuttle transit times, then they will have to pay very significant penalties", said the brokers' report. Target prices will apply to the half of the contractors' bill which relates to the digging of the tunnels themselves. Lump sum payments will cover the buildings at each end and tunnel fitting out whilst the locomotive supply and specialist engineering will be put out on competitive subcontract. A target price for the tunnels of £669 million plus FFr5864 million inclusive of fees was agreed for Equity 2. This figure is based on September 1985, prices but will be adjusted for inflation using construction industry standard formulae. Independent auditors retained by

Eurotunnel will then pay the contractors each month for the actual cost of the work they have compeleted, and will include a flat fee for profit.

The actual cost of the Tunnel plus fee will then be compared with the target. If the actual cost is less than the target, the contractors will split the saving equally with Eurotunnel. If the actual cost exceeds the target by up to 20 per cent, the contractors will have to pay 30 per cent of the overrun. But if the actual cost is more than 20 per cent over the target cost, everything beyond the first 20 per cent will be paid by Eurotunnel. Once actual costs exceed target costs, the contractors will not get any further profit payments.

Anything over budget during this phase of the building works would be horrendously expensive on two counts. It is not just the size of sums involved, which makes a small percentage fluctuation have a dramatic impact on the margin from which the equity investors' profits are due to come. Extra costs will also be compounded by extra interest charges. An overrun early in the tunnel excavation will have to be paid for with extra borrowed cash and the extra debt carried right the way through.

It would cost the contractors £75 million in penalties if they came in 20 per cent over budget. On the other hand they stand to make an extra profit of £62 million if they complete 10 per cent below the target. If the project did run into problems there would naturally be strong incentives for the contractors to get the targets changed in their favour. As they are responsible for both the design and the building, the opportunities for this to happen have been cut back.

But if Eurotunnel itself moved the goal posts by changing the specification, by demanding extra space for more rail tracks or wider tunnels, for instance, then it would have big problems in controlling the costs of the project. The British Ministry of Defence can testify as to how difficult it is to keep project costs from rocketing once suppliers are asked to incorporate a stream of changes into an order. Potential investors will have to decide for themselves if the chances of costs escalating in this key phase of the Tunnel project are likely to severely reduce the returns they have been promised. Eurotunnel does not have a big construction project track record, but its constituent companies, advisers and consultants are well experienced in large projects world wide. Eurotunnel argues with increasing force that the rewards available more than outweigh the risks.

Buildings and related infrastructure, the track, signalling and tunnel linings will all be built for a fixed cost agreed in advance. At Equity 2 the figures were £5568 million plus FFr5024 million adjusted for inflation. Any surplus on lump sum payments will make up the deficit if any element comes in over budget. The adjustment for inflation will be according to standard formulae, investors have been assured. If Eurotunnel decided to change any of the specifications, the figures would have to be recalculated and any extra payment would be worked out using a mechanism a bit like the one for the target prices on the actual Tunnel. The contractors would be paid their costs plus a percentage fee to cover design, overheads and profit. This time they would be getting a percentage of cost to cover profit, and not a flat fee, increasing the room for drift in the overall price, although probably not by very much.

The safeguard to ensure that Eurotunnel gets value for money from all of the "lump sum" parts of the contract is that 95 per cent of the items included will have to conform to the standards laid down by external organisations, such as the UK Department of Transport and the railway companies. Independent inspectors will be examining all of this to make sure the individual systems and structures conform to their own guidelines. The remaining 5 per cent in this category is the cost of actual buildings, and here Eurotunnel will do its own quality inspections.

The final 10 per cent of contracts will be placed after competitive tenders. These will cover the locomotives and the shuttle vehicles. The chances of cost overruns are theoretically quite high here, especially as the shuttles have to be designed from scratch, or the long distance trains which will have to be made comfortable for travellers. Transmanche-Link had already built full scale shuttle models in March 1987, so testing will have at least five years to identify problem areas. If these proved severe, Eurotunnel would have no option but to pay the extra design and development bills. The brokers working to sell Equity 3 argue that even if costs for all the competitive tender shopping list escalate by 20 per cent, the total cost of the project would only go up by about 2 per cent.

Target cost and lump sum works contain heavy enough financial incentives for the contractor to do everything they can to finish within budget. Contractors could be liable for payments up to a maximum of £75 million plus FFr877 million in damages for delay and would be levied at a rate of £164,000 plus FFr1,918,000 for every day's hold up for the first six months after the official completion date. After six months the rate would go up by 50 per cent a day.

Seven major assessment points have been built into the construction programme at regular intervals. If at any of these the contractors are behind schedule, they start paying penalties. They will only get this money back if they catch up later on. There are 30 additional check dates at which Eurotunnel could start to prod the contractors into greater efforts and remove lesser amounts of money if they did not respond to its liking. If nothing else, this should lead to the birth of a new sport on both sides of the Channel. There would be endless fun watching to see if the Tunnel is on schedule and waiting for the public humiliation of the contractors if it is late.

These milestones, or kilometre-roche in this new Euro age, will also provide a good basis on which the quoted share price of Eurotunnel can fluctuate. Long before it starts to trade as a going concern, Eurotunnel could become a volatile stock as new information about progress altered the market's perception of its health. Shuttle journey times are also crucial to the viability of the whole Tunnel. The contractors' brief is to provide a system capable of making the journey in 35 minutes. A special test will be run to see how fast the trip can be done, and penalties will be levied on a graduated scale up to a maximum of £40 million from the contractors if it does not come up to expectations. From the point of view of equity investors, journey times are vital. Overall profitability of the scheme is sensitive to changes in peak time through-put capacity in later years when demand has built up, but initially this will not be a major problem.

More pressingly, the big competitive advantage of speed that the Tunnel has over the other cross-Channel transport systems starts to be eaten away with longer crossing times, although the advantage for long distance travellers who do not have to change modes of transport can never be countered by ferries or airlines. A rail tunnel was a second best option for both governments, but a drive-through scheme will follow in its wake.

Chances of traffic delays before gaining a place on the shuttle are slim. A fan-shaped system of railway sidings will enable efficient loading while passenger-only trains go straight through the terminal at Cheriton into the Tunnel. Journey time is comparable with present hovercraft speeds, Mr James Sherwood of Sealink pointed out in an anti-Tunnel tract put out at the time of Equity 2, "Eurotunnel an assessment of the Assessment". He did not draw attention to weather delays, sea sickness, or collisions. For motorists the main attraction of the Tunnel will be that it offers a faster overall journey time because of the smooth uninterrupted flow of vehicles across the gap between the UK and French motorway systems.

"We believe a fundamental error has been made by the Eurotunnel analysts. They have assumed huge numbers of people are going to divert from North Sea and Western Channel sea routes which provide the shortest transit time, the lowest fares and the least over the road mileage, to go down to the south east corner of England in order to use an expensive fixed link which will dump them in the remote French countryside" said a Sealink report.

The criticism obviously struck a chord. Soon after Equity 2 was out of the way the Eurotunnel's consultants produced revised traffic forecasts which showed them gaining a slightly higher proportion of a somewhat reduced market. And "remote countryside" is contradicted in other anti-Channel statements where France is depicted almost as one continued industrial connurbation.

To gain a respectable market share the shuttles have to be perceived as quick, slick, smart and efficient. An extra few minutes on the time it takes to get from one end of the tunnel to the other is hardly going to make a massive difference on its own. All the same, at the time of Equity 2 the prospect of snarl-ups caused by slower than expected shuttles putting off some potential travellers worried some of the institutions. But again the opposition scored a home goal, with the community near Waterloo Station in London protesting that traffic was vastly underestimated. Eurotunnel has said that at peak times it should be able to run one of its 126 car shuttles every 12 minutes. The whole system has been designed so motorists will be able to turn up at the terminus and drive straight on without booking. Eurotunnel has a detailed technical analysis of how it would ensure that delays do not build up to levels likely to put motorists off using the crossing.

Boardroom Wrangles

When Mr Michael Julien quit as group financial director of the Midland Bank in July 1986 to become a full-time director of Eurotunnel the move was hailed as a major coup for the Tunnel company. His appointment was announced by Lord Pennock at the same time that he officially postponed Equity 2 until the autumn. It was a moment when Eurotunnel needed to be seen to be doing something right, and Julien was deliberately portrayed in the role of the experienced City cavalry, come to the rescue of the embattled Eurotunnel and there to make sure it had a safe journey on the quest for its £5 billion in loans and outstanding equity. Julien was highly regarded as a smooth financial fixer. He had just extracted Midland Bank from the disastrous purchase it made of the Californian Crocker Bank and had subsequently shaken up Midland's own international management. He had worked for Pennock before at BICC and had also sat on the board of its construction subsidiary, Balfour Beatty, which was itself a founder of the Tunnel consortium. More importantly, he could move easily among the institutions on whom a very urgent sales job was needed if the October placing was not to be a flop.

Significantly, as it turned out, at the time of Julien's arrival at Eurotunnel it was not made clear exactly what his job was, or where he was to fit into the hierarchy. The press assumed he would effectively be running most of the show, certainly on the UK side, at least until all of the equity cash was safely in and no-one, least of all Julien himself, disabused them. On the 26 September more than two months after the move was first mentioned, Eurotunnel unveiled a newly structured management team, overhauled to take on the final dash to the Equity 2 deadline at the end of October. Julien was deputy chief executive, with specific responsibilities for finance, planning and law. Very much what had been expected.

He seemed happy enough, was much in evidence urging sundry insurance company fund managers and the like to sign up for a chunk of the offer, and certainly there were no external signs of strife with the rest of the Eurotunnel management. The near disaster which Equity 2 turned into inevitably put considerable strains on organisation. But as late as 4 December 1986, five weeks after the Equity 2 deadline, Julien was reported in the accountant's trade magazine, *Accounting Age*, as saying he would increasingly be working with the construction companies on a day to day level, making sure the timing, spending and rate of development ran to plan. He also said he was looking forward to the political side of his job. As Julien must have known, it was not going to be that simple. The trauma of Equity 2 so nearly going off the rails had convinced the UK Government that changes in the team running Eurotunnel were vital. Pennock's position looked increasingly precarious.

The Bank of England moved even before Equity 2 was out of the way and insisted that Sir Nigel Broackes be co-opted onto the Eurotunnel board. Outside, it was assumed that he would soon be taking over from Lord Pennock. The plan was for there to be no official announcement of Broackes' arrival to be made before the issue was complete. In fact it was widely known among the fund managers before the issue closed that Broackes was again about to take an active part in the fixed link project. Even if his scheme had lost, Broackes had successfully raised £200 million for the Dartford crossing, and although this was admittedly a much smaller project, the image of the charismatic Trafalgar House boss was riding high in the City. It later became clear that the knowledge that he was about to enter Eurotunnel played a significant role in convincing key London institutions to take a stake in Equity 2. All the parties involved still insist that the timescale was very tight. Broackes said he never attended any meetings with institutions considering taking a stake in Equity 2. Broackes did, however, say that Michael Julien felt he had been deeply misled by Eurotunnel. Julien did not realise when he took the job that it was the intention to appoint another two deputy chief executives, one with responsibilty for construction and one with responsibility for operations. "Instead of being equal second he found himself equal fourth", said Broackes.

"When I joined in late October it was clear that for the first nine months after the contract was awarded no work had been done", said Broackes. "The government awarded the contract to an organisation that promptly disappeared". There had been a crop of departures after the mandate had been decided, most notably the chairman, Henderson, and the financial chief, Mr Quentin Morris. When Pennock took over, the priority had certainly been to see Equity 2 through to a successful conclusion. Broackes argues that this came to obsess the Tunnel company to the exclusion of all else. His own relationship with Pennock started to deteriorate rapidly.

"I wanted to know how the shuttle would work and what the customer appeal would be. I just thought there must be someone at Eurotunnel who knew more about it than I did. There wasn't", Broackes said. He came increasingly to despair about the organisation of the company. Trafalgar House also wanted its share of Eurotunnel contracts. The founder construction companies were not about to allow the outsider a place. As far as they were concerned, it had taken none of the risks, had been the most dangerous opposition and now was trying to elbow in where it was not welcome. The row finally came into the open at the start of February 1987. Broackes was considering resigning, principally because of the refusal of the other construction groups to let his company have a share of the work, but also because he was unhappy about the organisation of the whole project.

Rumours, that Pennock would stand down as chairman of the project grew much stronger. The consensus was that Broackes analysis of the organisational shortcomings of the company was largely correct, even if he was eventually prompted to leave for more partisan reasons. Pennock had additional problems being a director of Morgan Grenfell, the merchant bank increasinly being drawn to the centre of the Guinness affair. Eurotunnel officially denied that Pennock was to stand down, in an attempt to stop harmful rumours. The denials hardly had much credibility and themselves added to the toll of damage the company's image was taking.

For a while the row looked to be deadlocked. Eurotunnel refused to confirm that any changes of any sort were even being contemplated, while an increasingly unlikely group of aging industrial superstars were rumoured to be turning down the chairmanship, and the departure of Nigel Broackes was more and more widely expected. One problem was that the Eurotunnel official line was just what the City did not want to hear. The boost the Tunnel had been given in October with Broackes' appointment was largely because it was expected to herald changes in the leadership and the bringing in of new, or at least additional, financial advisers. If the only resignation was to be that of Broackes, it was likely that the maximum amount of damage possible would be done to the project's credibility. Things were not helped by an unfortunate piece of timing on Eurotunnel's part. It had chosen the new year to start a poster, newspaper spread and generally more up-beat public relations campaign. It was due to peak at Equity 3 in the summer but the increasingly violent ructions from the boardroom were making the early stages look silly.

According to Broackes, the temperature rose even higher in the boardroom when Pennock neglected to tell other Eurotunnel directors that he had decided to take up the executive chairmanship of Morgan Grenfell. The two jobs were hardly compatible nor was it really on for one man to find time for both. It has to be assured that by the time Pennock did tell the Eurotunnel board of his new role with the bank he had decided to step down at the Tunnel company. It was Pennock's predecessor as chairman, Henderson, who finally broke the official silence of his future. In an interview on BBC radio, he confirmed that Pennock would be going, but not immediately. The former ambassador officially designated the whole Tunnel company a lame duck until someone else could be found to take over. There followed an even more unseemly hawking of the chairmanship around an ever wider collection of possible candidates, with a considerably embellished tale told in the press.

It was increasingly clear that it was the Government which was every bit as keen as Broackes to see Pennock go. Commenting on Pennock's decision to stand down the *Financial Times* pointed out that there had been strong criticism of his leadership from both the Bank of England and Whitehall. Its Lex column added ruefully that the new chairman, whoever it was, would have an awful lot of hard work to do if Equity 3 was to stand any real chance. A week after Pennock started to shuffle off-stage, Broackes announced his resignation.

The contractors may have succeeded in rebuffing a rival from the top spot on what they regarded as their show but Broackes was due to have played a major part in Equity 3. Two days later, Julien delivered another hammer blow. He was going too. His new job was one of the hot seats at Guinness, by then busy rebuilding its own management structure. Julien said the project management and engineering skills needed to do the Eurotunnel job did not fit in with any of his own. The gulf between this and the earlier reports was enormous. The papers believed that to some extent he had just been given a better job offer and decided to take it, but the timing could not have been worse for Eurotunnel.

His loss was seen as the most damaging of the three resignations. For a start it had not been expected and secondly Julien was not supposed to have been deeply involved in the rows between Pennock and Broackes. Only a few days before, when the recriminations over Broackes' departure were still flying, Julien was being portrayed as one of the bedrocks of stability inside Eurotunnel. In a closer examination of the departure, the papers warned that his leaving deprived the company of its main British senior executive, all the rest being French. As he was also the money man, and as London was the international location where the money was inevitably going to be most difficult to get in time for Equity 3, from a UK perspective it could not have looked much blacker.

If the Tunnel was ever to be built, February 1987 had to be the low point of the company's fortunes. From then on things would have to be better. As an afterthought, Nigel Broackes argued that Eurotunnel suffered from what he calls the "grandee" syndrome — the assumption that conspicuously successful corporate businessmen can start up a major construction contract company and make it work. The French chairman, Andre Benard, is rightly highly regarded, said Broackes. He is conspicuously less complimentary about Pennock. Neither man understood how to build an efficient management team from scratch, he argued. Both have spent nearly all of their careers in big organisations and had never started anything outside the support structure big companies can provide.

Into this environment stepped Alastair Morton, the new UK cochairman. His first moves were to postpone Equity 3 from the summer to the autumn of 1987 and to bring in some heavyweight financial advice in the form of S. G. Warburg, the merchant banking group, to supplement the existing team. It ensured that his honeymoon lasted right the way up to the mini Equity 3 (called 3a), planned for the summer. Six months is a long time in the life of a big construction project yet to get properly underway: plenty of time to sort out the cost and revenue worries the banks and insurance companies have, and to get fully on top of the project management.

It is also just about long enough to spruce up the public relations to the point where some of the beneficiaries of the UK Government's south of England boom would consider risking a few hundred pounds on an investment in a piece of infrastructure, which, if all goes to plan, is going to make them still better off, just by being there. The benefits to the regions will follow in the wake of the southern wave, even producing ripples between Clydeside and North America. Equity 3 will show how well Morton has done.

Profits and returns for investors

Eurotunnel is expecting to turn in its first profit in 1993, the year the Tunnel is due to open for business. The *Pathfinder* prospectus, put out at the time of Equity 2, sets out the turnover, profit and dividend the company expects to come from running the Tunnel from 1994 to nearly the end of the 55 year concession.

Table 3   Eurotunnel projections of turnover, profit and dividends

Years ending 31 Dec	1994 £m	2002 £m	2012 £m	2022 £m	2032 £m	2041 £m
Turnover	752	1320	2608	4923	8955	15,150
Profit before taxation	109	727	1919	3896	7260	12,192
Profit for the year available for distrib.	65	433	1148	2329	4336	7214
Dividends — total	111	433	1148	2329	4179	7350
— per unit	3.34	11.82	31.34	63.59	114.10	200.68

On the basis of these figures, Eurotunnel then calculated the rates of return for the shareholders. It was keen to point out that the numbers it arrives at did not constitute a forecast and had been based on a host of assumptions, the correctness of which had to be a matter of judgement.

Project returns
Pre-tax return on project   16.7%
Post-tax return on project   14.1%
Post-tax return attributable to shareholders   17.7%

Dividend return	(net of tax)	(gross)
Return to subscribers of Equity 2	15.6%	17.2%
Return to subscribers of Equity 3	15.0%	16.6%

These figures apply to a UK resident shareholder, and would have to be adjusted further for those under a different tax regime.

The brokers appointed by Eurotunnel for the sale have been rather more adventurous in their projections. Using essentially the same assumptions and economic model, they have outlined the rates of return that shareholders climbing aboard each of the three stages of the equity programme can expect and what happens to these expectations if some of the key assumptions are varied. They looked at the impact of inflation being higher or lower than the model assumed; of the start being delayed; of the ferry companies pricing their rival services more aggressively than had been assumed, and of high speed trains running on the system. They also looked at the impact of withdrawal of duty-free concessions, as there is a chance that duty-free sales for travellers between EEC countries may be stopped before the Tunnel is complete.

High inflation during the construction period followed by a revival of sound money during the first years of operation of the service is the worst possible scenario for investors. The high inflation would put pressures on costs whilst the return of stable prices would make loans bite deep. As any mortgage owner will know, run-away inflation soon eats away the bulk of the heaviest debt. Inflation has only to fall to 4.5 per cent after 1993 and stay there, for the shareholders' overall return to drop by a full percentage point.

Eurotunnel will be pointing out to potential investors in Equity 3 that on the most likely assumptions they stand to make a clear 11.1 per cent return, if inflation is stripped out before doing the calculations. In any terms that is a healthy profit. Also the more pessimistic analysis has ignored the capital gain on the equity over the construction period.

The key construction marker comes in 1989 or 1990, early in the project, when the service tunnel is completed. Once that is done, major risks in construction are out of the way. All the areas where digging is difficult will have been identified and progress should be smooth all the way to the first shuttle.

If the traffic forecasts also prove to be correct, there should be nothing to stop the Tunnel meeting all the expectations its promoters have. No fixed link across a river or estuary operates under capacity; the chances are that Eurotunnel has undersold itself, but the remaining pieces of the Tunnel jigsaw will not complete the picture until 1993.

Table 4   Rates of return for investors when some of the assumptions change
(Source: *Eurotunnel Assessment*, Scrimgeour Vickers and Fielding Newson-Smith)

Case	Equity 1 + Share options	Equity 2 (figures in per cent)	Equity 3
Equity — Free Cash return to equity	25.4	17.8	16.6
— Dividend return to equity	21.7	16.4	15.4
Assumptions — HST + duty-free: GD + 2.15% p.a.(1983–93), + 2.75% p.a. (1993–2003). + 1.4% (2013–23), minimal thereafter: inflation at 3.5% in 1986, 5% in 1987, 5.5% in 1988 and 6% thereafter.			
High Inflation — Free cahs return to equity	27.9	19.8	18.7
— Dividend return to equity	24	18.2	18
Assumptions as in Equity case except for 7.5% inflation throughout			
Delayed Start Equity — Free cash return to equity	22.9	16.7	15.6
— Dividend return to equity	20.2	15.6	14.7
Assumptions as in Equity case but 1 year delay in start up.			
Low Inflation — Free cash return to equity	20.8	14.2	12.9
— Dividend return to equity	17.6	13.1	12.1
Assumption as in Equity case but with 3.5% inflation throughout.			
Without HST or Duty-free — Free cash return equity	22.6	16.1	15
— Dividend return to equity	10.3	14.9	13.9
Assumptions as in Equity case but no duty-free or HST.			
With HST, No Duty-free — Free cash return to equity	24.9	17.3	16.1
— Divided return to equity	20.9	15.9	14.9
Assumptions as in Equity case but with no duty-free sales.			
Aggressive Ferry pricing — Free cash return to equity	22.8	16.2	15.1
— Dividend return to equity	19.5	15	14.6
Assumptions as in Equity case but with 40% reduction in real return by ferry operators.			
High Growth Case — Free cash return to equity	27	18.9	17.7
— Dividend return to equity	13.2	17.5	16.5
Assumptions as in Equity case but 3.5% GDP growth throughout.			
Equity without contingency Provisions			
— Free cash return to equity	26.8	18	16.8
— Dividend cash return to equity	22.2	16.6	15.6
Assumptions as in Equity case but without cost contingency built into all the other cases.			

HST: high speed train.

2

Politics, Lobbying and Diplomacy

Nicholas Comfort

False starts and the choice of a project

The genesis of the current Channel Tunnel project has been anything but
smooth. There have been times when it appeared to have run out of
steam, usually because of difficulties over arranging the private finance
on which the British Government insisted, and others when the strength
of lobbying in Britain against any fixed link appeared likely to smother it.
The French Establishment must often have wondered whether it had
been wise to give Britain another chance after its unilateral withdrawal
from previous schemes, most recently in 1975. Indeed on 24 May 1983,
eighteen months after agreeing with President Mitterrand on the need to
press ahead with a fixed link, Mrs Thatcher was able to tell Dover Harbour
Board, on what was admittedly an electioneering visit to the port, that
construction of a tunnel was "not a live issue". For those committed to
such a project either as part of a broader European vision or as something
strictly practical, the frustrations over the interplay between the political
and diplomatic process, the lobbying and the financial hurdles have been

immense, with deadlines being missed, Government enthusiasm suddenly waning and the darkest hour coming just before the dawn.

The debate over whether a Channel Tunnel — or bridge — should be built has been conducted almost entirely in Britain. There has never been any serious argument in France over the case for a link; discussion has been confined to which scheme should be adopted and whether the British could be trusted to go through with it. Yet the lobbying in Britain against a fixed link has been able to draw not only on specific environmental and practical objections, but on a deep feeling in the English national psyche that any such scheme would end the emotional security of an island people. Lord Pennock, in his valedictory Lords' speech as Eurotunnel chairman, suggested that opposition to the tunnel had more to do with Hitler, Napoleon and Philip II of Spain than anything else [1]; others would reckon that the Huguenot tradition in Britain and the legacy of the Puritans was a factor — away from Kent, the strongest hostility monitored by Conservative MPs was in Oliver Cromwell's native Huntingdonshire. The lobbying against was at its most effective when the two governments were preparing to consider rival schemes at the close of 1985, rather than when the two Houses of Parliament were holding public hearings on the detail of the Channel Tunnel Bill. Without Mrs Thatcher's personal commitment to the idea of a fixed link at that critical time, the whole enterprise could have broken down in recriminations.

The political process which brought the present Channel Tunnel scheme within sight of fruition really began in February 1979 when British Rail and the SNCF, after two years of preparation, presented to the two governments an outline proposal for the "mousehole" — a single-track rail tunnel 5.6 metres in diameter through which "flights" of through passenger and freight trains would pass alternately in each direction. Then costed at £752 million (now nearer £2 billion), the "mousehole" was the cheapest conceivable crossing of the Strait of Dover, and also subject to neither of the objections which had ultimately killed the previous scheme. Its inability to handle "shuttle" trains ruled out the massive terminal on the Downs behind Folkstone which has always posed such strong environmental objections. And the absence of a high-speed rail link to London removed the environmental and financial doubts which had caused the hard-pressed Wilson Government to panic at the turn of 1974/75.

Given that the whole idea of a fixed link had so recently appeared buried, the promotion of the "mousehole" scheme provoked surprising interest within the contracting and financial communities. The French Government under President Giscard d'Estaing was clearly in favour of the principle, but the political situation in Britain was too fluid for it to have much faith in a joint venture. Within weeks of the report being

Margaret Thatcher and François Mitterand, outside Number 10, Downing Street, on
October 16, 1986.

submitted, the Labour Government, now headed by James Callaghan,
fell and was succeeded by Mrs Margaret Thatcher's Conservatives. The
new Prime Minister was known to dislike railways, took an abrasive line
with the rest of the European Community and was eager to cut public
spending; her support appeared unlikely. And public scepticism over the
proposal was reflected in a lack of serious lobbying against it, which also
owed much to the fact that even if built the "mousehole" did not appear
much of a threat to the ferry operators.

Nevertheless, things began to happen. A small Channel Tunnel Unit
was set up by the new Minister of Transport, Mr Norman Fowler, to
examine details of the scheme, as British Rail and the SNCF looked deeper
into the project after promising results from feasibility studies. In October
1979, Mr Fowler appointed Sir Alec Cairncross, the former head of the
Government Economic Service who had given crucial approval to the

previous scheme, to advise on the viability of this one; his sceptical report was to prove an equally serious damper. And on 19 March 1980, Mr Fowler made an announcement in the Commons which was to set the parameters from that point onward, not only establishing the need for any scheme to be privately financed — something that would have been inconceivable just a few years before — but leaving the inference that Mrs Thatcher's Government would give a viable project a fair wind. "The cost of any scheme would be very large," he told MPs, "and I should make clear now that the Government cannot contemplate finding expenditure on this scale from public funds. However, if a scheme is commercially sound, I see no reason why private risk capital should not be available." [2]. Mr Fowler went on to invite anyone interested in promoting a privately-financed Channel link, including British Rail, to submit detailed proposals.

The Commons Select Committee on Transport, chaired by the Labour (later SDP) MP Tom Bradley, a senior railway union official, immediately set up an inquiry into the possibilities. It considered a range of options including several put forward by contractors who were to be active later in the competing consortia; both the Eurotunnel scheme and the EuroRoute proposal for a bridge–tunnel–bridge link for rail and road were put to the committee in outline form. When it reported in March 1981 after taking evidence on both sides of the Channel, the committee opted for a single-track rail tunnel, but of 6.85 metre diameter, enough to take "shuttle" trains of cars and other vehicles when they could be fitted in. The cost was put at £806 million, and the committee accepted that this should be met by the private sector... though it felt the Government should help pay for any facilities included for a possible later expansion. Its readiness to endorse private funding followed Mr Fowler's disclosure when giving evidence that a "group of British merchant banks" was interested in providing finance; the Minister confessed himself "enthusiastically in favour" of a Channel link provided it were viable, as he believed it could be. The committee came close to sharing his enthusiasm, but told the Government it should produce a White Paper justifying its eventual decision and put it to Parliament for approval, and might well have then to hold a public inquiry into its implications [3].

The ball was now in the court of intending promoters, eight of whom had already been in touch with Mr Fowler. One was the European Channel Tunnel Group, in partnership with British Rail, which was ready to build either a single- or a double-track rail tunnel with "roll-on roll-off" facilities; it comprised the contractors Costain and Spie Batignolles with Dutch and West German partners; the Conservative MEP Sir David Nicolson and M Pierre Billecocq, a former French Transport Minister, as co-chairmen and Rothschild and S G Warburg as merchant bankers. An

Anglo Channel Tunnel Group was being put together; chaired by Sir Raymond (later Lord) Pennock, then chairman of the Confederation of British Industry, it was backed by British Insulated Callender's Cables, Balfour Beatty, Edmund Nuttall and Taylor Woodrow, with Morgan Grenfell handling finance. Channel Tunnel Developments 1981 comprised Tarmac, Wimpey and the merchant bankers Robert Fleming and Kleinwort Benson. Redpath Dorman Long were pioneering the bridge–tunnel–bridge precursor to EuroRoute, and advocates of a bridge, initially as a supplement to a rail tunnel, also mobilised to put forward three schemes. The combined effect was thoroughly to alarm shipping and port interests, which in July 1981 produced a report stating that any fixed link was unnecessary given the capacity of ferry services, and would be an economic disaster [4].

The stakes were raised again with the formal launch in September 1981 of EuroRoute, whose proposal for a bridge–tunnel–bridge for road traffic with vehicles spiralling underground from artificial islands, and a separate rail tunnel, was the most ambitious project with a price-tag at the time of £3.8 billion. It was headed by Mr Ian MacGregor, then chairman of the British Steel Corporation, with powerful industrial and financial backers including GEC, Barclays Bank, Trafalgar House (the construction group whose chairman, Sir Nigel Broackes, went on to chair EuroRoute and briefly serve on the Eurotunnel board) and the merchant bankers Kleinwort Benson.

At this point the French authorities began to believe that Britain might, after all, be serious about a fixed link and could be trusted to see it through. M. Billecocq of the ECTG had told the Select Committee that in the light of the "bad experience" of 1974/75, France felt it was up to the British Government to prove its good faith before any joint action could be taken. Informal discussions were under way by the time President Giscard d'Estaing was ousted by the Socialist M. Francois Mitterrand, with a Socialist–Communist coalition subsequently taking a clear majority in the National Assembly. The new President and his Transport Minister, the Communist M. Charles Fiterman, concluded that the British meant business, and M. Fiterman broached the subject with Mr Fowler early in September 1981. It was put on the agenda for the London "summit" meeting between Mrs Thatcher and M. Mitterrand the following week, and the two leaders wasted no time in committing themselves to agreeing on a scheme and embarking on it by 1984. Despite the interventionist Socialism of M. Mitterrand's first Government, he was ready to consider private financing. Officials in London and Paris opened talks within a month on the economic, technical, legal and administrative aspects of the project [5], and M. Fiterman headed off to Brussels to seek support from the European Commission.

Flexilink's poster campaign around London played on all the stereotyped prejudices.

Mrs Thatcher and M. Mitterrand caused considerable bafflement on both sides of the Channel by committing themselves so thoroughly to a venture which held out political and financial risks. There were reasons why neither leader might have felt warm toward the proposal, yet both were to stick to it despite a two-year hiatus when the financing was in doubt. At times each wanted a different scheme, with Mrs Thatcher attracted by the idea of a drive-through tunnel — she considered a shuttle train service for motorists no better than a ferry — and M. Mitterrand by a bridge [6]. But this apparently unlikely partnership was to hold, most importantly in the weeks, days and hours leading up to the decision to go for Eurotunnel. Mr Nicholas Ridley, Mrs Thatcher's Transport Secretary at that time, put her commitment down to "looking for monuments" [7]. Others maintained that she saw such a massive construction scheme as public proof of her concern for the jobless. Once open, the tunnel would in her view bring economic benefits to the whole of Britain, provided the business community took their chances. She may also have seen a privately-financed tunnel competing with the cross-Channel ferries as a blow against monopoly. There was certainly no fear at 10 Downing Street of Britain's sovereignty being compromised; there were, however, anxious questions from prominent members of the Royal Family. As for President Mitterrand, he too was motivated by the desire to see through a major project with which he would be associated. But his commitment was also nostalgic, stemming from the four years he spent in Britain during the Second World War. For France, as a Continental country, a cross-Channel link is both less important and less emotive than for Britain, but the improved communications it would open up — to West Germany and the Benelux countries as well as London and Paris — would provide a major stimulus for the depressed Pas de Calais region. In this M. Mitterrand was strongly backed by his first Prime Minister, M. Pierre Mauroy, a former President of the Nord-Pas de Calais regional council, who saw the tunnel as essential to future development there. But for M. Mitterrand there was little need to justify the project — it was simply the natural thing to do, provided the British could be trusted to see it through.

The mood of euphoria among advocates of a fixed link created by the "summit" of September 1981 and the accompanying flurry of activity among contractors and merchant banks was to be short-lived. On 21 October Mr David Howell, who had succeeded Mr Fowler as Transport Secretary, told MPs that Anglo-French consultations might take a little longer than expected [8]; a decision by the two governments on which scheme to adopt became scheduled for March 1982. Shortly before then Mr Howell told the Commons Transport Committee that a decision was in prospect although none of the schemes submitted entirely met the Government's criteria; he indicated that Britain would only support a bored rail tunnel and looked to the introduction of a Parliamentary Bill in the autumn.

It was not to happen. Serious doubts were emerging in Government over whether a fixed link would after all be viable. The Cairncross Report, submitted in November 1981 and published the following April, said that there was "no overwhelming case for a fixed link" and that it was doubtful whether investment on one would bring a better return than development of the cross-Channel ferries [9]. At the same time the Oxford University Transport Studies Unit, in a report commissioned by port and ferry interests, told Mr Howell that a tunnel would lose at least £144 million in the first 25 years, while a bridge might "just break even" [10]. At the end of April, the report of the joint Anglo-French technical working party landed on the desks of Ministers in London and Paris and, despite denials at the time, curbed the headlong rush. It ruled out anything as limited as the "mousehole", about which nothing further was heard, and concluded that building no link at all would be a realistic option [11]. Mrs Thatcher's Cabinet was preoccupied at this time with the darkest moments of the Falklands conflict and briefly let its interest flag. Mr Howell — an enthusiast for the tunnel but an ineffective Minister — did announce, on 16 June, that a group of British and French banks would carry out a further study, but it was now assumed nothing concrete would happen until after the next General Election. The anti-Tunnel element within the Cabinet had meanwhile been strengthened by the resignation over the Falklands of Lord Carrington, who as Foreign Secretary, had championed the project, and the ascendancy of the Chancellor of the Exchequer, Sir Geoffrey Howe, who made little secret of his hostility. (As Foreign Secretary Sir Geoffrey was to pursue a less critical line, the hostile line within the Treasury being inherited not by Mr Nigel Lawson, his successor as Chancellor, but first by Mr Peter Rees (MP for Dover and obliged by constituency feeling to be firmly opposed) and then Mr John MacGregor, who served in turn as Mr Lawson's deputies).

1983 turned out to be General Election year in Britain, with virtually nothing happening to further the project. Mrs Thatcher, as has been mentioned, visited Dover during the campaign and was lobbied hard against any challenge to the ferries. To Mr Rees' electoral advantage, she was able to say truthfully that things were off the boil. What was actually going on was a behind-the-scenes dispute between would-be promoters and the Government over the latter's refusal to give financial guarantees. Ministers argued that the scheme had to be entirely financed by the private sector, and that meant the private sector taking risks; the promoters retorted that as the previous scheme had been cancelled by the Government without warning, they should at least be indemnified against that. The dispute overshadowed the report from the banks on the financing of a fixed link, which, when submitted to the British and French Governments in May 1984, suggested that private enterprise could raise the money for a tunnel or a bridge, and that a twin-bore tunnel with "roll-on, roll-off" facilities could make a profit in its second year of operation. However, construction costs had virtually doubled since 1979, running well ahead of inflation.

By now the French were running short of patience, M. Fiterman telling members of the European Parliament in January 1984 that a decision to go ahead was needed by the end of June [12]. The fortunes of the tunnellers reached their lowest ebb when, on 22 May, Mr Nicholas Ridley, who had become Transport Secretary the previous October, rejected the banks' recommendation for "marginal guarantees" for a privately-financed link. Mr Ridley said in a written House of Commons answer: "it has been and remains the Government's firm position that any project would have to be financed entirely without the assistance of public funds and without commercial guarantees by the Government." [13]. The gulf between British Ministers — notably Mr Ridley, one of Mrs Thatcher's most enthusiastic disciples on economic policy — and what had formed itself in March 1984 into the Channel Tunnel Group appeared unbridgeable; the French, who shared some of Mr Ridley's misgivings, issued a "wait and see" statement. Any follow-up between the two countries was ruled out by the collapse of the Socialist–Communist coalition in France and the formation of a new Socialist Government under M. Laurent Fabius.

Hoverspeed, argued Tunnel critics, could cross the Channel as quickly as the shuttle.
Price, safety and weather conditions were not as openly discussed.

While the summer of 1984 seemed the leanest time for the advocates of a Channel Tunnel since the previous project was abandoned, work was going on behind the scenes which was to transform the situation. The newly-formed Channel Tunnel Group (CTG), an amalgamation of previous consortia around what became the Eurotunnel scheme, was redoing its sums in the knowledge that even if the Government gave a guarantee against political cancellation, the economics of the project on current assumptions were against it. By early autumn, CTG was able to tell Mr Ridley it could cut the construction period from six to four and a half years through new boring methods and round-the-clock shift working, raising the direct costs of the project by 10 per cent but shortening the period before investors would see a return. CTG's new arithmetic was completed just as Anglo-French talks resumed in earnest. Officials met in Paris in mid-October, and on 14 November, Mr Ridley agreed with two French Ministers, M. Jean Auroux (Transport) and M. Paul Quilès (Planning) on a private-sector link, with Government guarantees solely against cancellation on political grounds. On 29 November, Mrs Thatcher and M. Mitterrand endorsed the agreement at a "summit" meeting in Paris, following which the Prime Minister, in the words of Sir Nicholas Henderson, the former British Ambassador to France who was about to take the chair of the CTG, "did not make a matter-of-fact statement of support; she enthused" [14]. However, she cautioned the French public that it would be a "very, very long time" before a Channel Tunnel were built [15].

From 30 November 1984 to 28 February 1985, six British and six French officials worked against the clock to produce specifications for whatever link were built, for example that it must be adequately safeguarded against terrorist attack, should last for at least 120 years, have a minimum speed of 80 kilometres per hour and enable travellers to be rescued from any point in an emergency within 90 minutes. Mr Ridley and M. Quilès approved the draft as it stood, though with the French Minister inserting a caveat about the need for checks against rabies in order to calm British fears on the issue [16]. And on 21 March 1985, the two governments gave promoters until the end of October to submit detailed schemes, in the hope of reaching a decision by 31 January 1986.

The formal "Invitation to promoters" was issued by the two governments on 2 April 1985, with Mr Ridley making it clear to MPs that he was now firmly committed to this "exciting and imaginative project" [17], and that the chances of a fixed link being built now lay squarely with the promoters. He also gave a first indication that once a scheme had been selected by the two governments, it would be dealt with in Britain by way of a hybrid Parliamentary Bill and not through a public inquiry — a statement which brought immediate protests from Labour. But he assured MPs that the guidelines would "ensure that there is adequate public consultation, that environmental, social and employment impacts are fully appreciated and that the financial conditions are fully met" with the two governments considering all of these points before reaching their decision. The 64-page "Invitation" covered every aspect of tunnel planning, financing, construction and operation, requiring the promoters to produce a tunnel or bridge scheme within the parameters set, and submit it in time to the civil servants in charge of the process: Mr Andrew Lyall at the Department of Transport in London and M. Raoul Rudeau of the Ministere de l'Urbanisme, du Logement et des Transports in Paris.

The "invitation"

(1) Spelt out the financial guarantee against cancellation for political reasons while repeating that no others would be given;
(2) Promised the British Government would use its "best endeavours" to get the necessary legislation through;
(3) Explained the French political procedure;
(4) Outlined road and rail improvements which would be necessary on either side of the Channel;
(5) Laid down basics for accident and emergency services;

(6) Committed both nations to the speediest possible customs and immigration checks;

(7) Cautioned that any bridge piers or artificial islands would have to confirm with maritime safety rules;

(8) Required promoters to supply financial "evidence of robustness and viability", explaining how large a sum would be raised and when;

(9) Set out details of what was to be built, the construction timetable and breakdowns of cost;

(10) Called for promoters to carry out environmental assessments in both countries; and

(11) Demanded safeguards against rabies, plant disease, terrorism and sabotage.

The political argument between the various groups of hopefuls — notably CTG and the higher-profile EuroRoute — had been joined as soon as it became clear in November 1984 that the fixed link was "on" again. Quite apart from those MPs who came to speak for particular consortia of promoters, or for groups outside opposing a fixed link, clear battle-lines could be seen with supporters of a tunnel or bridge far outnumbering committed opponents, who were nevertheless more vocal. It was easier to identify the antis than the pros: half-a-dozen Conservative MPs from Kent who felt the scheme would destroy the local economy or environment — notably Mr Jonathan Aitken of Thanet South — or were obliged to say so in order to be re-elected; a slightly larger number of committed Tory anti-Marketeers who saw the project as binding Britain closer to Europe; the Scottish Nationalist leader Donald Stewart for similar reasons; a knot of North-Eastern and Scottish Labour MPs who feared the project would benefit the South-East at the expense of their own regions; committed Labour anti-Marketeers (though not all of them) and Labour MPs sponsored by trade unions most alarmed by the effect of the project on their members (notably seamen and dockers). Remarkably it was December 1986, when the Tunnel Bill was almost through the Commons, before an all-party anti-tunnel group was formed, by the former Labour Minister the late Mr John Silkin. There had been a pro-Tunnel group, though not a very active one, since before the cancellation of the previous scheme; unbeknown to its officers it had financial backing from the interests that eventually comprised Eurotunnel [18].

Labour itself had continual difficulty in deciding a policy line on the Tunnel, given that rail and some construction unions were strongly in favour, and maritime and road transport unions generally against, with local Labour parties in the Dover area vociferously opposed. What usually happened was that when conference motions on transport were "composited" in advance of debate, reference to the Tunnel was omitted. Mr Neil Kinnock, Labour leader from 1983, was never totally opposed,

Francis Bouygues with a model of the shuttle to go through the Tunnel. His seal of approval encouraged other contractors to join France Manche, the French half of Eurotunnel. Source: Bouygues.

but generally set conditions (such as the demand for a public inquiry) which would make it unlikely the link would ever be built. Eventually, early in 1987, the party's national executive adopted a "facing both ways" policy; if it came to power before the Bill had received Royal Assent, it would call a public inquiry, while if work had begun it would increase the tunnel's potential by "dispersing" more traffic to the regions. The SDP/ Liberal Alliance by contrast was in favour, though the Portsmouth MP Mr Mike Hancock with constituency ferry interests was opposed.

    Among the governing Conservatives, pro-Tunnel elements were in a clear but generally latent majority. The opponents had for the most part campaigned against membership of the EEC, and in many cases were to be ardent opponents of the Anglo-Irish agreement of November 1985. There was also a leavening of ideological Right-wingers such as the

"Selsdon Group", and opponents of the railways — notably Sir Alfred Sherman, a former speech-writer to Mrs Thatcher. A formal Conservatives Against the Tunnel group was eventually set up, its patrons being Mr Aitken, whose opposition was based on his constituency's dependence on ferry-based tourism and the anti-European legacy of his grandfather Lord Beaverbrook; two Midlands MPs, Mr Patrick Cormack and Sir John Farr, and the fervently anti-Market Mr Teddy Taylor, MP for Southend East. Their influence at Westminster was limited, but they reflected strong feeling in some constituencies and their denunciations of Eurotunnel's financial prospects added to the the consortium's headaches over raising capital.

A clear sign that the "Chunnel" was now a runner was the launching by Sealink British Ferries — the company previously owned by British Rail which had just been acquired by the Bermuda-based Sea Containers group of the American magnate James Sherwood, re-creator of the Orient Express — of a vigorous campaign of hostility. Mr Sherwood's opening shot was to warn that if the link were built, ferry operators would pull out of Dover and possibly Folkestone, turning them into "ghost towns" [19].

Sealink was also co-founder of the most strident anti-Tunnel group, not to mention the one which caused the most irritation to Tunnel promoters — Flexilink. Formed at the end of April 1985, Flexilink drew its strength from Sealink and European Ferries (trading as Townsend Thoresen and now merged with P&O's ferry interests), and the Dover Harbour Board, a veteran of previous anti-Tunnel campaigns through its role in handling over 60 per cent of cross-Channel traffic, plus Hoverspeed (only later owned by Sealink) and a number of port operators and Chambers of Commerce. The ferry operators argued at the Flexilink launch that a tunnel was unnecessary given their own ability to provide a service, and would put up to 40,000 sea- and land-based jobs at risk in East Kent; Mr Jonathan Sloggett, chairman of Flexilink and managing director of the DHB, declared that all the schemes put forward demonstrated "a total ignorance of the requirements of this demanding market" [20]. Ferry operators and maritime unions also argued in the wake of the Falklands that continuation of cross-Channel services was essential to provide transport for troops in time of war. Flexilink entered the battle at a time when the promoters were too preoccupied with preparing their own schemes and lobbying the decision-makers to launch a frontal attack on public opinion; indeed the CTG and EuroRoute concluded a publicity truce to save each other time and money (21). Furthermore, the "anti" campaign began with the public far from convinced of the need for a fixed link; a Gallup poll that summer showed 50 per cent in favour, 37 per cent against and 13 per cent undecided, compared with 69 per cent in favour and just 17 per cent opposed in 1963. The poll also indicated a strong preference for a tunnel among those who did want a link of some sort: 43 per cent for a tunnel, 9 per cent for a bridge and 17 per cent for a combination of the two [22]. Flexilink's campaigning, articulated by Mr Sloggett, Sealink's James Sherwood and behind the scenes by Sea Containers' director of communications Maureen Tomison, went straight for the jugular, harnessing the widespread English distrust of anything foreign in support of its triple aims of persuading the Government to reject all the schemes submitted, blocking the successful project in Parliament and through the courts, and sabotaging fund-raising in London financial markets ... with the last proving the easiest of the three, given the risks involved.

At the start of September 1985, with just eight weeks to the deadline for promoters, Flexilink launched a £400,000 poster advertising campaign "for cross-Channel choice". The Tunnel promoters questioned whether it had won many hearts, but by the time Eurotunnel was officially chosen the following January, Gallup was recording a 51–36 majority against the project [23]. The sophisticated end of Flexilink's argument was an assessment from the stockbrokers Phillips and Drew questioning the financial viability of a fixed link and raising the spectre of a completed bridge or tunnel going bust and putting the ferries' finances, in turn, in jeopardy. But the poster campaign, which *The Economist* described as "giving off a nasty smell" [24], conjured up vivid pictures of French juggernauts thundering through a tunnel populated by implicitly-rabid rats, with the caption "The Channel Tunnel — the black hole that will put Britain in the red." Another advertisement carried the slogan: "There's something about this project that smells — and it isn't just garlic."

The likelihood of a Channel link being built, with inevitable disruption to the environment not to mention the effect on ferry jobs, was also prompting an upsurge of protest in Kent: in the Dover/Folkestone area, in the unemployment blackspot of Thanet to the north, and in communities along the routes from London to the Channel ports who feared it would make life even more unpleasant for them. An Anti-Channel Tunnel Society (ACTS), decribed by its founder, Mrs Kathy Methven, as "a Christian organisation to protect our cultural heritage" was set up in Folkestone, and within a year claimed 25,000 supporters [25]. When Mr David Mitchell, Mr Ridley's deputy, went on a three-day "swing" through Kent to assess public reaction to the prospect of a fixed link, he met a polite but frosty reception; he antagonised townspeople in Dover by describing opponents of any link as "little men and little Englanders". And CTG made enemies by pulling out of a public meeting in Dover on 19 October — attended by EuroRoute — on the impact of the rival schemes. But until a particular scheme was chosen, the protests were fragmented, and at no time did a single opposition movement on environmental

David Mitchell of the the UK Department of Transport visited the old Channel Tunnel workings that were abandoned in Janary 1975, ten years later on August 19, 1985. Source: Murray Hughes

grounds take shape, each community preferring to present its own case and organisations such as the Council for the Protection of Rural England giving an overview.

There was opposition, too, on the French side of the Channel, though never as widespread or as forceful as in Kent. The Pas de Calais as a whole might benefit from the construction of new European transport links through it — indeed some British MPs feared that if built, the Tunnel would suck jobs through from England - but the ports of Calais, Dunkerque and Boulogne had a great deal to lose. Chambers of

Commerce and trade unions had, since the close of 1981, been voicing concern not only that ferry and port workers could lose their jobs, but that the sizeable tourist and shopping trade built up for day-trippers by sea would be lost. The Social and Economic Committee of the Nord-Pas de Calais regional assembly concluded in May 1985 that the implications for the French Channel ports were "alarming in every sense" [26]. Boulogne would probably lose nearly all its ferry traffic, not to mention its hoverport. Calais could lose 5000 jobs from its passenger ferry terminal alone, with M. Guy Flamengt, director of the port's Chamber of Commerce, describing the overall effect as "an economic disaster in an area that lives off the sea" and promising outright opposition to whatever scheme were proceeded with. However, the French Government showed no more willingness to preserve the ferry services than it had been to support indefinitely the region's coal, steel and textile industries, whose decline made the impetus offered by a Channel link so attractive. And by September 1986, M. Jean-Jacques Barthe, Communist Mayor of Calais and the town's MP, was lamenting that while his local supporters remained "extremely hostile" to the project, his surgery was now filled with men anxious to get construction jobs when work began [27]. Nord-Pas de Calais was also benefiting from EEC assistance in a way depressed areas of Thanet and the Kent coalfield were not — a document produced during the Select Committee hearings showed that while neighbouring areas of France had received £28 million from Brussels, Kent had received just £80,000 [28].

It was against this background that the rival promoters prepared their submissions to the two governments. Promoters based in Britain hurried to find French partners to make their schemes more acceptable — CTG joined forces with France-Manche on 1 July — and those discovering that their plans fell foul of the guidelines in the "invitation" hastened to modify them. All with serious hopes tried to read the minds of Ministers, and staged subtle or not-so-subtle lobbying campaigns, with that in London having the higher profile. The CTG showed a ten-minute video to MPs in the basement of a Westminster hotel, and at the political party conferences that autumn, CTG and EuroRoute took stands within hailing distance of each other to sell their own proposals and discreetly rubbish their rivals. The general assumption was that the final choice would rest between CTG's update of the rail tunnel scheme of the 1970s, and EuroRoute's ambitious bridge–tunnel–bridge proposal; CTG believed EuroRoute, now headed by Sir Nigel Broackes, which had been lobbying hard for three years or more, had "peaked" too soon [29]. There was also Eurobridge, the consortium headed by Lord Layton, which proposed a four-level, twelve-lane road suspension bridge with spans up to 5 kilometres in length, and a single bored 6-metre-diameter rail tunnel; this did its sums but never had more than an outside chance.

Barely a month before the 31 October deadline, James Sherwood threw a spanner in the works — and nonplussed his Flexilink partners — by coming up with a scheme of his own and turning the full force of his personality and political contacts on to promoting it. His Channel Expressway in its original form provided for a road tunnel with rails set in the floor, through which trains could pass every so often once the roadway had been "swept" clear of traffic. Newspaper articles began to appear stating that the scheme was a runner; it was never clear whether these reflected an interest among politicians or created it, but in a matter of weeks considerable support was there. Mr Sherwood was eventually persuaded that using the same tunnels for road and rail traffic was not on, and he revised the scheme to include separate tunnels for both, but by that time the 31 October deadline had passed and his competitors were crying "foul". Cynics were quick to suggest that Mr Sherwood might be seeking to establish a bargaining position rather than develop a fixed link himself; there were also complaints from Mr Sherwood that when he bid for Sealink, he had not been made aware that the British and French Governments were on the point of agreeing to a form of competition which would completely alter the basis of its finances. This argument found few listeners, although Mr Sherwood had actually bid for Sealink when the prospects for a tunnel were looking fairly bleak; however he had told shareholders in a letter that Sealink was in the best position to challenge a fixed link.

The deadline for the submission of plans duly expired on 31 October, amid scenes of pantomime involving Sir Nicholas Henderson, whose meeting with Mr Ridley was held up because security staff at the Department of Transport had apparently never heard of their Secretary of State [30]. While the British heads of consortia were meeting Mr Ridley and posing for pictures, their French counterparts were going through a similar ritual in Paris. When the dust had settled, it turned out that nine schemes had been submitted, five of which were immediately rejected as non-starters [31]. Ranging from the ingenious to the downright batty, they were:

* The Bournemouth-based Boothroyd Airship Co's proposal for a suspension bridge held in place by heavy-lift airships at an estimated cost of under £1 billion. Mr Milton Boothroyd claimed that six 400-tonne airships could bear a double-deck bridge with three lanes of traffic in each direction and a trolleybus shuttle service.
* Euro-Trans World Channel Tunnels, designed by Mr Eric Munday of London. This rail-only tunnel was priced by its inventor at £115 billion; he did not have financial backers for it.

* A road/barrage project from a Mr M L McCulloch;
* Eurolink, a road bridge proposed by a London design group; and
* Twin rail tunnels submitted by a Mr M Stoodley.

This left four serious contenders: Channel Expressway's twin-bore tunnels (at this stage still taking both road and rail traffic); CTG's twin-bore rail tunnel; EuroRoute's combined bridge and tunnel for road and rail; and Eurobridge, the outsider. And the battle between the promoters was now joined in earnest as the two Governments assessed their schemes and strove to agree on one of them.

On 5 December the House of Commons Select Committee on Transport declared its support for the CTG's rail tunnel scheme, on the casting vote of its chairman, the Labour MP and former railway union official Mr Gordon Bagier. This narrow squeak for the ultimately successful project heartened rival promoters, though Mr Bagier insisted that he had only needed to intervene because MPs known to favour the CTG scheme were unable to attend. The committee backed the project because it used proven technology, should be cheaper to build than its rivals and would run fewer risks of overshooting its timetable or budget. It rejected arguments for a road link, but added that if the Government considered one "indispensable" it should plump for EuroRoute [32]. At the same time Sealink issued the results of a MORI poll showing almost twice as much public support for Channel Expressway as for any other scheme. And the *Sunday Telegraph* reported that Ministers were "now believed to be leaning heavily in favour" of Mr Sherwood's project [33]. The paper also said that a Countryside Commission survey had concluded that all the projects would have a major impact on the Kent environment, but that those involving a bridge would be the worst, having a particularly serious effect on the white cliffs of Dover.

The next Monday, 9 December, the Commons debated the issue, with a procedural motion to adjourn the House being carried by 277 votes to 181, a majority of 96. The main argument was over the Government's intention to promote the chosen scheme by Parliamentary Bill rather than by a public inquiry, which both sides acknowledged could take years and jeopardise the promoters' chances of raising the money. Mr Kinnock had earlier told reporters that while most Labour MPs probably favoured a fixed link and he personally supported a rail tunnel, the Government was taking a decision without consulting the public [34]. Rejecting Labour's demands, Mr Ridley noted that it had taken 17 years to get a decision on a new by-pass for the West Country town of Okehampton and 22 years on the expansion of Stansted Airport, adding: "Delay has become a weapon used under our planning procedures in order to frustrate development. I make no apology for saying that we must proceed at reasonable speed." [35].

Salvage work on the ill-fated *Herald of Free Enterprise* in Spring, 1987. The disaster also damaged Flexilink's campaign against Eurotunnel which had made great play on the comparative safety of ferries.

Next day there was a furore in the Commons over a television report that a sizable number of MPs were being paid as "consultants" by the promoters of rival schemes or by opponents of a fixed link, and were not declaring their interests when speaking on the subject. One particular charge was that Mr Den Dover, a Conservative MP whose vote had been decisive in the Transport Committee's preference for CTG, was being paid £8500 a year by one of its constituent companies. The Labour MP Mr Geoffrey Lofthouse spoke of conduct "bordering on corruption" [36], and Mr David Winnick, another Labour member, asked: "How can an MP speak and vote when he is paid a financial reward by a company?" The Speaker, Mr Bernard Weatherill, told the House that "every MP knows if he has an enterprise it should be registered. It is a matter of honour." Peers connected with companies competing for the franchise took the hint and in that Friday's debate on the proposal, fell over each other to declare an interest. That weekend the *Observer* reported that 37 peers and MPs, mostly Conservatives, had business ties with companies involved with the consortia or their opponents [37] — though this did not mean they personally supported a particular scheme or had a direct financial interest. CTG's muster included the Conservative MPs Sir John Page, Tim Smith, Robin Squire (also Parliamentary adviser to Sea Containers) and John Ward, and eleven peers: the Conservatives Lord Bellwin, Viscount Hood, Lord Boardman, Viscount Sandon, Lord Crawford and Balcarres and Lord Pennock, the Alliance peers Lord Taylor of Gryfe and Lord Ezra, and three independents, Lord Balfour of Burleigh, Lord Boyne and Lord O'Brien. Those connected with backers of EuroRoute included the Conservative MP Sir Frederic Bennett and ten peers, Lord Nelson of Stafford, Lord Limerick and Lord Rockley (Conservative), Lord Gregson (Labour), Lord Chandos (SDP), Lords Richardson, Weinstock and Ashton of Hyde (Independents) and the non-aligned Lords Catto and Matthews. Eurobridge could muster three peers including its originator, the Liberal Lord Layton. As far as the antis were concerned, Mr David Atkinson, Conservative MP for Bournemouth East, raised some eyebrows as adviser to Grayling, Flexilink's public relations company; Labour's Lord Mulley, probably the only former Transport Minister to line up against a fixed link, was working for Sealink and Mr Squire (again) was advising Sea Containers.

The announcement of the successful scheme by Mrs Thatcher and M. Mitterrand was set for 20 January 1986, in Lille. The British side would have liked longer and suspected that French Ministers wanted an early announcement to give them as much time as possible to take credit before the National Assembly elections in March, which the Socialists expected to lose [38]. CTG, EuroRoute and Channel Expressway were all still in the frame, though CTG suspected that Mr Ridley was showing an interest in the Sherwood scheme to find out the point at which the various consortia might be prepared to co-operate with others. The French wanted to eliminate Channel Expressway at the outset because of the late changes made to it and the lack of French participants, and Mr Ridley's reluctance to drop it was seen in Paris as a sign of British commitment [39]. French sources have since described the British as having defended Channel Expressway with "desperation" [40]. (A MORI poll commissioned at this time by Sea Containers suggested that a majority in the Pas de Calais wanted Channel Expressway if a fixed link had to be constructed [41].) On 3 January 1986, Mrs Thatcher chaired a Cabinet committee which put the schemes in order of preference, and four days later Mr Ridley and M. Auroux met in Paris to thin out the field. The Ministers speedily agreed to drop the Eurobridge proposal, but when M. Auroux suggested eliminating Channel Expressway as well, Mr Ridley argued that it had important advantages, particularly drive-through (which appealed to Mrs Thatcher though she had doubts over its practicability). The British Minister added that his Government did not rate EuroRoute, doubting whether the combination of bridges and a tube on the seabed was feasible and reckoning the cost excessive. M. Auroux, in turn, insisted that EuroRoute stay in; it had strong backing in France, not least within the Élysée. French banking interests had lobbied hard for EuroRoute, reckoning CTG's financial forecasts "too good to be true", and they took the project's eventual rejection very badly, accusing M. Auroux of having been biased against it because of his previous connections with SNCF [42]. The CTG scheme was barely mentioned at the Ridley–Auroux meeting [43]. However, lobbying by the influential contractor Francis Bouygues, an original bridge supporter who had joined the France-Manche consortium although he considered the Tunnel project somewhat unimaginative, was tilting the French Government in its direction [44].

The immediate sequel to the meeting was that CTG told Mr Ridley it was ready to incorporate a commitment to the eventual construction of a road tunnel in its scheme; Mr Ridley briefly took this as suggesting that CTG was ready to join forces with Mr Sherwood, but eventually the point was taken and it proved crucial in swinging the decision to CTG. In the days that followed, Mr Ridley met the chairmen of all three consortia still in the running and told Mr Sherwood he was in a strong position but must not say a word; the Sealink owner promptly went out and told a Press conference that he had as good as won. The two Ministers met again in London on 13 January; they came close to ruling out EuroRoute, partly on security grounds, but its abandonment had to be part of a larger package. Sir Nicholas Henderson states that a French emissary visited Sir Robert Armstrong, the British Cabinet Secretary, to tell him that although President Mitterrand favoured EuroRoute, he would probably switch to CTG if Britain dropped its interest in Channel Expressway [45]. The French Government believed acceptance of the Sherwood scheme would be seen in the forthcoming election as "surrender" to British interests [46]; it also suspected that Mr Sherwood aimed to build his tunnel with cheap labour from South Korea [37]. Sir Nigel Broackes partly understood what was happening, and launched an attack on Mr Ridley for "extraordinary partiality" toward Channel Expressway [47]. By the time this appeared, Mr Ridley and M. Auroux had had a further meeting which clarified the situation but did not finally resolve it. The day before the choice was to be announced, there was Press speculation that CTG had won. But the consortium agreed the formula for its drive-though commitment with officials less than 24 hours before the announcement, there were last-minute problems over an agreement between CTG and the French Railways, and even on the day, Sir Nicholas could not be certain CTG had won, though he thought it probable [48].

Mrs Thatcher and M. Mitterrand arrived at Lille Town Hall on 20 January to the strains of "Marlborough s'en va-t-en guerre", to announce that the Group had been awarded the concession to construct a twin rail tunnel capable of carrying through traffic and "shuttle" trains, and had given an undertaking to submit plans for a drive-through link by the year 2000. A treaty would now be concluded between Britain and France enabling the work to take place; it would need to be ratified by

People travel with their cars on rail shuttles in several long Alpine tunnels; there have been no casualties in 40 years of use. Eurotunnel used the system as an example of how safe its own system would be. Anti-Tunnel groups argued that the systems were not comparable.

Parliament and the National Assembly before work could begin. The two leaders looked to completion of the project as "a landmark in the development of the relations between the United Kingdom and France and Europe as a whole" and as providing a vital link in the European transport network. Mrs Thatcher was moved, during a brief speech in French, to acclaim the challenge of the Tunnel as "passionant." In the Commons, Mr Ridley had a subdued but generally friendly reception as he explained the decision. He made much of the opportunities the scheme and the services stemming from it would offer to British industry, and assured anxious Kent Conservatives that there was no question this time of a high-speed rail link through the county.

A White Paper published on 4 February justified the choice of the CTG proposal on the ground that it

"offers the best prospect of attracting the necessary finance";

"carries the fewest technical risks that might prevent it from proceeding to completion";

"is the safest project from the traveller's point of view";

"presents no problems to maritime traffic in the Channel";

"is the one that is least vulnerable to sabotage and terrorist action";
"has an environmental impact that can be contained and limited" [49].

The announcement in Lille was also followed, in France, by the launch of a major package of job creation measures for the Nord-Pas de Calais region in advance of the Parliamentary elections and, at Westminster, by a Commons' debate which produced a vote in favour of 268 to 107, a majority of 161. Five Conservative MPs, four of them from Kent, voted against the Government. In a separate vote, a Labour demand for a public inquiry was defeated by 263 to 173, a majority of 90.

On 12 February 1986, Mrs Thatcher and M. Mitterrand met again, this time in the medieval Chapter House of Canterbury Cathedral, to initial with their Foreign Ministers the treaty binding both countries to the project. Demonstrators against the project — including a spirited group of French seamen — were kept at a distance, so Canterbury City Council sent a letter of objection over the choice of venue which prompted a vigorous reply from Mrs Thatcher. The Prime Minister spoke of the event as opening a new chapter of industrial and business links between Britain and France; the President less prosaically quoted Joan of Arc and T. S. Eliot, adding in a frank reminder that previous agreements had not stuck: "There is no going back on this. The Channel Tunnel will become to schoolchildren part of the geological scenery of our planet." [50]

Getting the show on the road

With the CTG scheme finally chosen, a race against time began for the consortium to secure Parliamentary approval — though in France that was close to a formality — and raise the funds necessary for construction. Opponents of any fixed link were now joined by local amenity groups who knew for the first time precisely what they were up against. Those with the greatest concern were residents of hamlets just outside Folkestone who were confronted with plans for a massive "roll-on, roll-off" terminal, and neighbourhood activists around Waterloo Station on London's South Bank who found it earmarked as the terminal for "Chunnel" trains; Mr Ridley was not alone in wondering whether the French would have opted for a rail tunnel had they known that trains from it would be directed to a station named after Napoleon's final defeat. CTG, for its part, knew that the two governments were now committed to giving it a fair wind ... and took some time to realise that the consortium could be its own worst enemy.

Although the two governments had now made their commitment, there was still one formality to go through: a "concession agreement" with the successful consortium, under which the two governments agreed to acquire the necessary land and lease it to Eurotunnel, initially for 55 years. This was signed on 14 March by Mr Ridley, M. Auroux, Sir Nicholas Henderson and M. Jean-Paul Parayre, chairman of France-Manche, and the British and French partners, while retaining separate identities, promptly formed themselves into Eurotunnel. March also saw the defeat of the Socialists in the National Assembly elections, which gave rise to what the French characteristically termed "cohabitation": a Socialist President, M. Mitterrand, coexisting with a Right-wing Prime Minister, M. Jacques Chirac. James Sherwood had claimed prior to the elections that M. Chirac favoured Channel Expressway and would repudiate the agreement reached, and there were some weeks of unease on the British side before Mr Ridley — himself about to move on to the Department of the Environment — could ascertain that the French commitment to what was now Eurotunnel stood. Even a year later there were still rumours, encouraged by some former French supporters of EuroRoute, that M. Chirac, though firmly committed to some sort of fixed link, would prefer another scheme.

The argument in Britain was now to centre on Westminster. Despite the occasional debate, the odd Select Committee inquiry and sporadic questions from MPs and peers, Parliament had so far had only a walk-on role; now it was to take its full share of responsibility. The Department of Transport took the lead in promoting with Eurotunnel a Bill approving details of the scheme and ratifying the Treaty with France which would be put through both Houses of Parliament in turn, with the Royal Assent (a formality at the close of the process) being necessary before the consortium could issue shares. The Bill would be a "hybrid", a special type of measure used when a matter of public policy directly affects private interests. A similar Bill had been promoted for the scheme killed in 1975, and had passed the Commons and was about to begin its progress through the Lords when the Government abandoned it. "Hybridity" had in the meantime briefly been a political issue when the Labour Government promoted legislation to nationalise the aircraft and shipbuilding industries, only for the Speaker to rule that it would have to be dealt with by the lengthier Hybrid Bill procedure because it affected private interests; Ministers sidestepped the issue by leaving out the ship repairing firms over which the dispute had arisen.

The peculiarities of a Hybrid Bill are twofold. In the first place, as a result of what Mr Ridley once referred to as its "private parts", it need not lapse with a General Election, as Government legislation would. A new Parliament can agree to pick it up where the previous one left off, sparing the promoters the need to start again from scratch. At the same time, one Parliament cannot bind the next, so there would be no guarantee that a Bill of this kind caught by the calling of an election would in fact be carried

The Channel tunnel.
Heineken submits its plans.

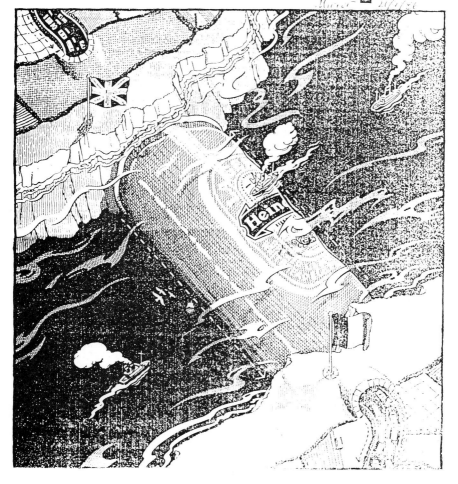

Each time the Tunnel hit the headlines, a new set of advertisers sought to capitalise on the popular theme. Heineken, the lager with the famous slogan of 'refreshes the parts other beers cannot reach' decided to refresh the choice of fixed link with its own proposal.

over. This was a live issue as the General Election came into prospect, given that a Labour Government elected with the Bill in mid-stage would halt the process while a public inquiry was held. The other peculiarity is that a Hybrid Bill may be considered by a Select Committee of each House immediately after its Second Reading and before it goes into Standing Committee for line-by-line consideration; such a committee is bound to hear petitions from anyone directly affected by the provisions of the Bill. In Mr Ridley's view this procedure amounted to "consultation, but national democracy" [51], combining the need for Parliament to decide whether the scheme went ahead with the right of the public to have its say.

Unfortunately for the Government and Eurotunnel, such opponents as Flexilink, the Labour Party and local amenity groups did not see it that way, reckoning the procedure a "carve-up" and no substitute for a full public inquiry. Dover Chamber of Commerce and other opponents also threatened to seek a declaration from the European Court of Human Rights that compulsory purchase orders made through such a process were invalid [52]. Another misfortune was that the Bill appeared just as backbench Labour MPs were fighting a guerrilla campaign against the use of Private or Hybrid Bills to promote major projects, such as the expansion of Felixstowe Docks — outside the Dock Labour Scheme controlled tightly by the unions — which they saw as having a major impact on the environment and the regional economy. Ministers argued that the procedure had covered the construction of the railway network in the 19th century before anyone had heard of public inquiries; their critics responded that if an inquiry were needed to build a road or a power station, it was illogical to exempt a railway, dock or tunnel. The issue was eventually shunted off to a Committee on Procedure, but not before it had cast the Channel Tunnel legislation in a poor light.

The Channel Tunnel Bill, as tabled, comprised 48 clauses and six schedules in 118 pages. As well as giving the consortium authority to build and operate the Tunnel, and ratifying the Anglo-French treaty, it set out particulars of the scheme (including new construction at Waterloo and upgrading of the A20 around Folkestone), authority for compulsory purchase and the regime under which the Tunnel would be operated. Private and hybrid legislation normally has to be introduced in November, at the start of a new Parliamentary session, unless a waiver is granted. The Department of Transport, putting its Bill forward in April 1986, thought this would be a formality, but the Commons Standing Orders Committee, chaired by the Deputy Speaker, Mr Harold Walker, decided on 6 May to take evidence from objectors to the scheme as to why the Bill should not go forward at once [53]. Had permission been refused, the promoters would have lost at a stroke six of the twelve months they reckoned they could afford before the "time-window" for approving, financing and starting work on the Tunnel had gone. The committee heard arguments for haste from the Department of Transport, and for delay from Dover Harbour Board, Sealink and Mr Jonathan Aitken, and on 21 May divided 5–5 on whether to grant the waiver, thus leaving a final decision on the matter to the House as a whole. On 3 June the Commons voted by 283 to 87, a Government majority of 196, to allow consideration of the Bill to begin; Mr David Mitchell, Minister of State for Transport, who had been chairing a working party on the environmental impact of the scheme with Kent local authorities, won over some critics by agreeing to extend the deadline for petitions against the project by ten days to 27 June.

The time allowed for consideration of the Bill was in itself a source of controversy. Mr Ridley estimated when the measure was published in April 1986 that it could be on the Statute Book by Easter 1987; that point was to be reached with it barely halfway through the Lords, having cleared the Commons. Opponents of the scheme asserted from the outset that it was being pushed through with "undue haste", citing the jumping of the queue for private legislation, the tightness of the timetable and procedures set by the Commons Select Committee, and the Government's last-minute reminder to objectors that they could petition the Lords, as evidence. Even the routine estimate from Government business managers prior to the Second Reading in the Lords that "this may take five•to six hours" was seen by some as proving the whole procedure a charade. However supporters of the project argued that there was plenty of time for full consideration of the scheme and its implications, and that critics of the timetable were actually complaining that the thing was happening at all. At no point was either side ever likely to accept that the other had a case.

The Channel Tunnel Bill came before the Commons for a Second Reading on 5 June 1986, amid controversy over a completely extraneous matter. The Labour MP Mr Tam Dalyell had secured time the following day for a motion attacking Mrs Thatcher's conduct in the Westland Helicopters' affair, and her readiness to allow American bases in Britain to be used for President Reagan's bombing of Libya. Conservatives set out to thwart him by staging an all-night "filibuster" over the Tunnel to wipe out the next day's business, but abandoned the idea at the last minute when it became clear the Opposition might take revenge by delaying essential legislation. The Second Reading, moved by Mr John Moore, who had succeeded Mr Ridley as Transport Secretary, was given with little argument, by 309 votes to 44, a Government majority of 265. Seven Conservatives and one Social Democrat voted against it; while the Labour leadership abstained, nearly 40 Labour, Scots and Left-wingers voted against, and one of the party's MPs, Mr Dale Campbell-Savours, backed

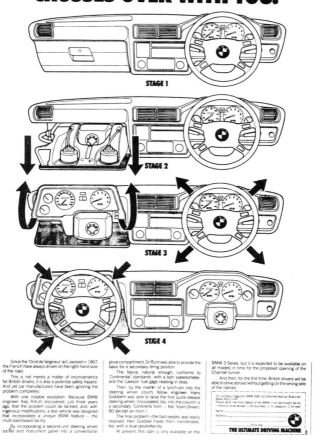

WHEN YOU CROSS THE CHANNEL, OUR STEERING WHEEL CROSSES OVER WITH YOU.

STAGE 1

STAGE 2

STAGE 3

STAGE 4

Since the 'Droit de Seigneur' act, passed in 1867, the French have always driven on the right-hand side of the road.

This is not merely a matter of inconvenience for British drivers, it is also a potential safety hazard. And yet car manufacturers have been ignoring the problem completely.

With one notable exception. Because BMW engineer Aap Riffuhl discovered, just three years ago, that the problem could be tackled, and, with ingenious modifications, a test vehicle was designed that incorporated a unique BMW feature – the multi-dashboard facility.

By incorporating a second-unit steering wheel socket and instrument panel into a conventional glove compartment, Dr Ruhl was able to provide the basis for a secondary driving position.

The fascia, naturaly enough, conforms to 'Continental' standards with a kph speedometer, and the 'Lawson' fuel gage reading in litres.

Then, by the insertin of a lynch-pin into the steering wheel colum, fellow engineer Hans Grabbem was able to dvise the first quick-release steering wheel (incorpated, too, into the column is a secondary 'Continenti horn – the 'Vonn-Drivers' 80 decibel air-horn.)

The final problem, r the foot pedals, was easily resolved. Herr Grabbei made them transferable, too, with a dual positiofacility.

At present, this opin s only available on the

BMW 3 Series, but it is expected to be available on all models in time for the proposed opening of the Channel tunnel.

And then, for the first time, British drivers will be able to drive abroad without getting on the wrong side of the natives.

To: Lise Adjurl Egound, BMW (GB) Ltd, Elkinletti Avenue, Bracknell, Berks RG12 4TA.
Please send me more details of the BMW multi-dashboard facility.
I intend to drive abroad ☐ for business ☐ for pleasure ☐ for ever.
Name _____
Address _____
_____
Post code _____

THE ULTIMATE DRIVING MACHINE

April Fools' day seemed an appropriate time for the West German car manufacturers BMW to suggest a novel way of dealing with driving on both sides of the road...by allowing a steering wheel to be changed from one side to the other, mid-Channel.

the Tunnel. A Labour amendment regretting that the scheme did not guarantee enough work for British firms, complaining that British Rail had not done enough to plan through rail services to destinations other than London and seeking even-handedness between the Tunnel and ferries had earlier been defeated. Lengthy argument followed over terms of reference for the Select Committee which was to hear objections, with Kent MPs under constituency pressure particularly anxious that petitioners should be fully heard. None was more active behind the scenes than Mr Michael Howard, Conservative MP for Folkestone and Hythe, who as a junior Trade Minister was unable to present his constituents' case in public. Recognising the strength of local anxiety over a scheme which was bound to wreak havoc and the ability of his Liberal opponents to exploit it, Mr Howard got the deadline for objections extended and also negotiated with Eurotunnel a special compensation scheme for payments of up to £2000 to householders in the hamlets of Newington, Peene and Frogholt, which were hardest hit by the proposed Tunnel terminal and the road and rail access to it.

Eurotunnel's readiness to set up this compensation scheme and to buy out householders in the hamlets at what it considered market rates in advance of the scheme gaining Parliamentary approval, was one of a number of efforts made by the consortium to demonstrate its ability to be humane and sensitive. Eurotunnel also held a programme of meetings in the area to be most affected and sent round a display in a caravan, and in January 1986 opened an advice centre in Folkestone. Such efforts were met with a degree less ridicule than the Department of Transport's own "public consultation exercise", amounting to little more than a rack of leaflets in a waiting room at its Westminster headquarters. Eurotunnel was foiled in its attempts to woo the public as a whole by an Independent Television Authority ban on pro-Tunnel television advertising until the scheme had been passed by Parliament. In the event, even newspaper advertisements backing the scheme did not appear until the following January, as the Bill was starting its passage through the Lords. And it was the opponents of the Tunnel who had the initiative as Parliament began hearing objections to it.

The nine members of the Commons Select Committee were picked by the all-party Committee of Selection; the SDP/Liberal Alliance passed up the opportunity to have a representative on a body which was likely to have to sit through much of the summer, and thus five Conservative and four Labour MPs were chosen. They were: Mr Michael Fallon (C, Darlington); Mr Alex Fletcher (C, Edinburgh Central); Mr Kenneth Hargreaves (C, Hyndburn); Mr Robert Jackson (C, Wantage); Mr Terry Lewis (Lab, Worsley); Mr Nick Raynsford (Lab, Fulham); Mr Allan Rogers (Lab, Rhondda); Mr Peter Snape (Lab, West Bromwich East); and Mr Lewis Stevens (C, Nuneaton). Unlike the Lords' Committee which was to hear objections the following year, there was no bar on members with strong views on the project or interests in its success taking part. In Mr

Snape's case this provoked immediate controversy; a front-bench Labour transport spokesman, he was not only a former railway signalman, sponsored by the National Union of Railwaymen, but an unashamed supporter of the Tunnel. His choice prompted a complaint in the Commons from Mr Jonathan Aitken, but Mr John Biffen, Leader of the House, rejected it, saying that all committee members were trusted to behave "honourably" [54].

On 19 June the committee elected as chairman Mr Fletcher, a 56-year-old former Scottish Office and Trade Minister, who with 13 years in the Commons was its senior member (most of the others had been first elected in 1983 and Mr Raynsford had only been in the House a matter of weeks, though he was to prove anything but a novice). Though not a political heavyweight, Mr Fletcher had gained a reputation for competence while at the Scottish Office and was a sound Parliamentary performer; his firm, if genial, chairing of the Select Committee antagonised some objectors, but he emerged from the process with a knighthood in the 1987 New Year's Honours. The committee decided to take evidence in public at Westminster three days a week until the end of July and again in the autumn, and also to travel to the Folkestone area in September to hear local objections. And it set strict limits to the evidence it would hear, which Mr Fletcher spelt out to the inaugural public session: "The principle of a fixed Channel link has already been given approval by Parliament. The Committee is not therefore concerned with arguments against the principle, but with its effect on certain private interests."[55] Once the implications of the ruling had sunk in, there were cries of "foul" from opponents of the scheme. Those cries grew louder, prompting threats of legal action from Mr Aitken [56], when Mr Fletcher made a further statement on 2 July after the full scale of the objections was realised. "Some people who are not directly affected by the proposals in the Bill have deposited petitions giving their personal views on the principle of the decision which was taken by the House of Commons on Second Reading," said the chairman. "Such people cannot reasonably expect to be heard. Despite the very large number of petitions, the committee will ensure that the legitimate interests of those directly affected by the proposals are given a fair hearing. Clearly this will be most sensibly done when a large number of people petitioning on similar matters are represented by a spokesman of their choice.... We are well aware of our duty to hear cases where individuals are directly affected by the Bill, but this does not mean that the Committee will feel obliged to listen to the same arguments over and over again."[57].

The Select Committee held its first public session in a packed Grand Committee Room at the Commons on 24 June amid scenes of black farce rather than controversy. A few minutes into his opening speech, Mr

Michael Fitzgerald QC, counsel for the Government, collapsed and an attendant was summoned with a stretcher to apply first aid. The

Newspaper cartoonists probably came up with the best fixed link suggestions, providing a use for the ferries and quick completion of a link. (Reproduced from the *Daily Mirror*, 9.12.1985, by permission of Syndicates International.)

attendant trapped his finger in the framework of the stretcher and began to bleed profusely. An ambulance was then called, only for the paramedics to assume that the attendant was the casualty; it took some time to explain that it was Mr Fitzgerald who was most in need of attention. The hearing was suspended for 20 minutes until order could be restored, with another member of the Government's array of "silks" completing the speech. Mr Fitzgerald turned out not to have been as ill as was at first feared, and was back in the room after the lunch adjournment.

Between that first public session and the production of its "interim report" on 5 November, the committee heard 220 hours of evidence covering more than 1900 pages in 36 days of sittings, dealing with 155 objectors in four days at Hythe alone. A total of 4852 petitions of varying degrees of relevance were eventually lodged against the Bill, from local authorities, ferry operators and major business and amenity organisations with concerns about the scheme — "objection" did not in law necessarily amount to opposition but gave the petitioner "locus standi" to address the committee — to individuals who stood to lose their home. The committee's task was to decide whether the Bill should be amended before it was considered line-by-line in Standing Committee; it eventually proposed 70 amendments, most of them minor and in their effect well short of what would have satisfied most objectors. Even though Mr

Fletcher ruled that barristers need not appear in wigs and gowns, the atmosphere of the Palace of Westminster intimidated some witnesses, a few of whom went home after seeing notices on the main door reading: "Closed to the Public". Committee members were sometimes called away on other Parliamentary business, and occasionally offended objectors by using the sittings to catch up on their mail; on occasion witnesses paused to ask: "Are you listening?", or rebuked the panel for yawning. Much of the discussion concerned whether undertakings by the promoters should be given statutory force. A number of issues were resolved while the MPs were hearing evidence, with lawyers for Eurotunnel, the Department of Transport and objectors closeted in an adjoining room, trying to settle a contentious issue which might otherwise take days. Eurotunnel, however, had a credibility problem, as objectors did not always believe the undertakings or place much weight on them, particularly the promise to "do everything reasonable and practicable" to ensure that minestone and other materials were brought to the construction site by rail [58].

The MPs found themselves looking for suitable sites for hang-glider enthusiasts who stood to lose their sloping field above Folkestone and for displaced sports' clubs, considering the fate of the great green bush cricket and a colony of rare newts, and adjudicating between rival schemes for road access to the Cheriton terminal which would have a major impact on the hamlet of Newington, and determining whether particular works were rightly included in the Bill. This last argument applied both to the plans for the Waterloo terminal and to the extension of the A20 round Folkestone toward Dover. Amenity groups argued that the road scheme should be dealt with under normal planning procedures and that its inclusion in the Bill was an abuse, the more so as the Tunnel would actually mean less traffic for the new section of road. However, witnesses from Dover argued that the slump the Tunnel would bring made the road project even more essential. There was similar conflict over the siting of the international railway station for Kent at Ashford rather than Dollands Moor, outside Folkestone; interests from the town wanted it at Dollands Moor, where sidings were planned in any case, to give easy access for Folkestone passengers to tunnel trains, people living near the site did not want the station or the sidings, and Ashford council wanted the station there. The committee had a first chance to see the sites in question when members visited Kent on 1 July; they had to contend not with demonstrators against the Tunnel but with somewhat basic travel arrangements provided on the spot by the Department of Transport, against which they rebelled. They paid a further visit to the most controversial locations while in Kent to hear evidence.

Lay objectors appearing before the committee were surprisingly fluent and succinct, but there were times when witnesses had to be brought sharply to order for dealing with matters outside the committee's remit. It was ten days before Mr Fletcher refused to hear an objector, largely because most of the early evidence came from Kent County Council and other local authorities concerned about details of the scheme but not opposed to it as such. All that changed when, on 16 September, the committee reassembled after a six-week break, in the palmy surroundings of the Imperial Hotel at Hythe to hear local objections. During four days there and two at Dover, some 4000 petitioners were due to be dealt with, and Mr Fletcher had to stress that their written arguments would be fully considered whether they testified or not. The hearings at Hythe were marred by arguments between the committee, witnesses and their representatives over how so many petitions should be handled. Even before proceedings began, Mr Aitken had written to the chairman that the timetable was "an outrage and totally inconsistent with the committee's promised duty to act fairly and reasonably towards petitioners." Mr Raynsford publicly endorsed this view and was told by Mr Fletcher that he had been a "naughty boy" to break ranks [59]. One witness broke off a business trip to America but failed to get a hearing before he had to return; others accused the committee of causing them unnecessary pain; some subsequently said they had received a fair hearing. The MPs heard a number of sad stories, none more so than that of the Frys of the Coach House, Newington, who with the plan as it stood would have a railway viaduct outside their bedroom window, and under Shepway Council's revised access plans would lose their home altogether [60]. At the other end of the scale, a Mrs Litten from Saltwood spoke of "this Royal throne of kings, this scepter'd isle" and was only ruled out of order when she tried to read a letter from the Queen Mother [61]. Brigadier John Mackenzie of Slaybrook Hall told the committee that Eurotunnel would succeed where Hitler had failed and that the tunnel would be "Allah's gift to the terrorists" [62].

Nor was the atmosphere at Hythe and Dover improved by the committee's refusal to hear evidence on the risk of rabies, terrorism, the import of drugs or the transport of nuclear waste. Mr Fletcher stressed that these were matters of public policy, not for the Select Committee but for the Standing Committee which was to follow. Some objectors, however, managed to say a good deal on the subject including a woman whose petition concluded: "I've put up with being attached to Wales, but the thought of being attached to the French is beyond belief. The thought of rabies, terrorism and the French, God preserve us from it." [63]. Other witnesses voiced fears of being attacked by rabid bats and concern that armed French policemen might roam through Kent, but they were more than outweighed by a wealth of serious and deeply-felt worries about the fate of local wildlife and flowers, possible local rail service cuts once the

Tunnel was open, and the damage caused if spoil from the Tunnel were dumped in the wrong place.

"The ticket office was closed at Upminster!"

Futuristic perspective that may not be so far from reality. An Englishman wearing a bowler hat is even more unusual though than a Frenchman wearing a Breton shirt! (Reproduced from the *London Evening Standard*, 22.1.1986, by courtesy of Jak.)

There were rebukes both at Westminster and in Kent for the ferry operators and Eurotunnel for trying to "rubbish" each other's financial credibility, and Mr Sherwood — who was fighting the scheme having failed to convince the two governments to choose his own — ran up against the groundrules set for the inquiry, at one point being told by the chair that he was wasting the time of the committee [64]. His presence was not seen by all anti-Tunnel campaigners as a boon; one agent for employees of other ferry companies asserted that "ferries are like toys to him" [65].

Nevertheless the ferry operators did contrive in the end to set much of the agenda. Both Mr John Drinkwater, QC, for Eurotunnel, and counsel for the Government attempted to have them ruled out of order, Mr Fitzgerald condemning their case as an "assault on the principles of the Bill" [66]. Most notably they were to get written into the Bill an undertaking that no Government subsidy could be paid to the Tunnel operators — thus meeting their concern that if the Tunnel opened and got into financial difficulties, it would either be bailed out by the State or leave the ferries with obligations they could not meet. The ferry operators themselves alternated between saying the Tunnel would put them out of business and declaring that they could take on the competition and beat it; the National Union of Seamen somewhat gave the game away by suggesting that Eurotunnel and the ferries should "divide up the market" between them [67]. It was the ferry operators also who first raised the issue of fire safety in the tunnel; if they could oblige passengers and cars to be segregated in tunnel "shuttles", they could adversely affect the relative economics of tunnel and ferry travel. (At the time, prior to the Zeebrugge disaster and the ensuing disclosures that lorry drivers often stayed with their vehicles, the ferry operators' evidence that segregation always took place at sea was accepted. Ironically, when Townsend Thoresen's Herald of Free Enterprise went down, Maureen Tomison of Sea Containers was in Switzerland preparing a video on the hidden dangers of car-ferry trains through Alpine tunnels.)

The ferry operators also faced charges from Mr Snape that they had "orchestrated" objections in an effort to "delay and discredit the project until after a General Election" [68]. At Westminster he drew from Mr Gwyn Prosser, an executive member of the ships' officers' union NUMAST and a highly-proficient presenter of petitioners' cases, the admission that Sealink had given him paid leave to represent objectors, many of whom were not employees of the firm [69]. And at Hythe, Mr Snape provoked shouts from the audience of "how dare you?" when he said after an argument with Mr Tony Neumann, another NUMAST official, that Sealink had "done their best to drum up support against the principle of the Bill over which this committee has no standing" [70]. Sealink, which had been handing out petition forms to travellers on its ferries, agreed that it had paid the £2 petition fee for many objectors not on its payroll, but rejected claims that it had even advised petitioners what to wear when they gave evidence.

The Select Committee reassembled at the Commons on 14 October as Eurotunnel's difficulties over raising its "Equity 2" financing package were at their height. Mr Aitken was drawing Ministers' attention to the consortium's struggle and inferring that the project faced imminent disaster; the Government was all too aware of the embarrassment failure would cause and was arm-twisting financial institutions with the help of the Bank of England. Mr Robin Leigh-Pemberton, Governor of the Bank, told a somewhat peeved Mr Sherwood that it had "done what it could to ensure that what have seemed transitory problems of organisation and presentation did not disrupt the Equity 2 operation", but was doing no more than offer advice when asked, as it would for any company. Sir

Nigel Broackes, former head of EuroRoute, was also persuaded to join the board and played a key role in raising the money; he was to part company with the consortium early in 1987 after its contractor-backers resisted his efforts to gain a slice of the action for Trafalgar House. In the Commons Mrs Thatcher, pressed by Tunnel supporters to renew her commitment to the project, responded that while there could be "no Government money", both Britain and France hoped such a challenging project would go ahead [71]. Eventually the £206 million was raised, leading the French to conclude that "Maggie has won" [72].

On 22 October, the Select Committee turned to the scheme's impact on London and, in particular, Waterloo. British Rail, which for the previous scheme had proposed a London terminal at White City and later favoured Victoria, now put forward Waterloo as the obvious destination for the 70 per cent of international rail passengers bound for the capital. The choice was resisted by a combination of the Left-wing Lambeth council and groups representing a vigorous community a stone's throw from Westminster Bridge. Their case was that expansion of the station and the extra traffic through it would put an unbearable burden on local roads and Underground services, give rise to litter, fast food shops and prostitution, create noise and danger for local people, put severe pressure on parking and cost 211 jobs as small businesses were moved out to make way for the terminal. Arguing that expansion at Waterloo had nothing to do with the Tunnel as such [73], they called for a public inquiry into the suitability of other terminal sites, with a combination of other existing rail terminals, Olympia, the former Bricklayers' Arms goods station and sites in Docklands suggested. The idea was also raised — to be pursued further in the Lords — that tunnel trains could use the Snow Hill link between Blackfriars and Farringdon which was then being reconstructed to provide cross-London local services. British Rail had argued that Waterloo with the Chunnel traffic would be handling fewer passengers than at the peak of commuting in the 1960s, and that some 550 new jobs would be created there; Councillor Bob Colenutt, chairman of Lambeth Council Planning Committee, disputed both the figure and whether the jobs would go to local people [74]. Other witnesses said bluntly that they did not want to be part of central London, and a series of local residents spoke of a "community fighting back" against a tide of office development, and now put under threat by the choice of Waterloo [75]. Mr Snape was to put the counter-argument with force: "It is somewhat illogical for some people living around Waterloo to say that their lives would be made untenable by approximately three extra trains an hour at a terminal which, if my memory serves me right, has been used as a railway terminus since 1848" [76]. When it became clear that the committee were not ready to exclude Waterloo from the Bill, the Waterloo Community

Seven Lords a leaping...almost. The Peers on the House of Lords Select Committee visited
the 1974 Channel Tunnel site in March 1987.

Development Group issued a bitter statement declaring that the petitioners had been "gagged, hassled, rushed and censored" by a "kan-
garoo court".

The toughest questioning in the Select Committee of how thoroughly
the scheme had been thought out came from the genial but persistent Mr
Raynsford. He asked few questions in the early stages but gradually
became a dominant figure, as his doubts grew over Eurotunnel's
competence to carry out the operation and over environmental and safety
aspects of the scheme. He effectively took charge for the first time on 31
July when Major Charles Rose, the Government's chief inspecting officer
of railways, was giving evidence on fire safety in the Tunnel [77]. But
when the committee began formulating its conclusions, he came into his
own. For while most of the committee were broadly happy with the plan
as it stood, Mr Raynsford was not. On 4 November the Commons voted
174–73 to let the Select Committee carry its full report over to the new
session, and the next day Mr Raynsford issued a "minority report",
declaring himself in "fundamental disagreement" on four points: safety
in the Tunnel, especially allowing passengers to remain in their cars on
"shuttle" trains, the access route to the Cheriton terminal (the Committee
on Mr Fletcher's casting vote rejected the 'Shepway Alternative" which

would avoid splitting Newington and Peene but demolish houses of note), and the failure to put the A20 extension and the Waterloo scheme to public inquiries.

The Committee's official report indeed proposed few changes of substance. The freight inspection sidings at Dollands Moor should be screened by trees, Eurotunnel should lower the "obtrusive" rail viaduct planned for Newington by half a metre, the stipulations that no Government funding should be provided and that cross-Channel ferries should be equally free of subsidy should be written into the Bill, and the Director-General of Fair Trading should review the pricing of all cross-Channel services before the Tunnel came into use [78]. The committee grudgingly left in the Bill a provision which would allow 3.75 million cubic metres of spoil to be dumped at the foot of Shakespeare Cliff, Dover, but called for an early review of possible alternatives. On 22 November, after the new session had begun, the committee published its full report which included a warning that through trains to destinations in Britain apart from London would not be feasible if the Customs and Excise continued to resist the idea of conducting checks on board trains [79]. Once again, the report was matched by a dissenting statement from Mr Raynsford claiming that the committee had failed to respond to public concerns.

The Bill now moved to a Standing Committee of the Commons, which was a very different animal. Made up of 19 MPs — 11 Conservative, seven Labour and a Liberal — it operated more on party lines, the Tory ranks including two Ministers (one of them Mr David Mitchell, who as deputy to Mr John Moore would pilot the Bill through committee) and a whip. It included three Labour MPs, Mr Raynsford, Mr Snape and Mr Lewis, who had served on the Select Committee, but a new cast of Tories, among them Mr Aitken. In the chair was Miss Betty Boothroyd, a senior figure in the Labour Party and an elegant woman who had once danced with the famed Tiller Girls troupe. She supervised the proceedings with such good nature that at the close Mr Aitken chivalrously presented her with a bottle of perfume..., "Channel No. 5". The whole ethos of a Standing Committee is incomprehensible to outsiders; amendments are proposed by the Government, to tidy up the Bill, by the Opposition or individual members, some are selected by the chair for debate and the ensuing discussion leads into all manner of digression before a vote is taken.

The Standing Committee met 15 times to consider the small print of the Bill between 2 December 1986 and the following 22 January. And from the outset, Mr Aitken and Mr Raynsford made it clear they would challenge most aspects of the scheme as planned; Mr Mitchell was moved to say that if the MP for Fulham were indeed not opposed to the Tunnel, as he claimed, he would hate to meet someone who actually was; his "scaremongering language" over safety and readiness to repeat himself also earned a series of rebukes from both sides of the committee [80]. Mr Snape, as Labour's official representative on the committee, was duty-bound to harry the Government, but as a supporter of the scheme (despite doubts over Eurotunnel's competence) was anxious not to wreck it.

One major debate was over the dumping of spoil below Shakespeare Cliff, to form a 32-hectare platform. Mr Raynsford's efforts to limit dumping to 1.85 million cubic metres failed, but the feeling persisted that Eurotunnel and Kent local authorities had not pursued the alternatives hard enough. The most obvious alternative, apart from dumping at sea, was Lappel Bank near the Isle of Sheppey in the Thames Estuary; Mr Mitchell pointed out that spoil could only reach the island over a railway swing bridge which in summer had to open up to 16 times a day for yachts [81]; the site was also close to a pharmaceutical factory. Greater dramas were to come, though; on 11 December Mr Raynsford claimed that "surplus French Chunnel spoil" would be dumped in England after M. Étienne Schwarczer, Directeur-Delégué of France-Manche, was reported as saying that three million cubic metres would be extracted at the French end and five million at the British [82]. Mr Raynsford declared that "the canny French have pulled a fast one", but Mr Mitchell insisted that there was bound to be an imbalance as geological conditions were worse on the French side and British tunnellers would have to dig 3.2 kilometres beyond the undersea boundary. Before the Lords considered the Bill, Kent County Council came back with a study ruling out alternative dumping sites to Shakespeare Cliff on environmental grounds.

The Standing Committee had always been billed as the forum for debating issues such as rabies, terrorism and security... with safety giving rise to the most crucial debate, either side of the Christmas and New Year recess. Mr Raynsford put forward the fundamental proposal that the safety authority being set up for the Tunnel should prepare regulations based on the segregation of "shuttle" passengers from their cars. The debate was conducted on the assumption that all passengers were segregated from their vehicles on ferries [83]. Mr Raynsford argued on the basis of evidence collected by the ferry operators and the Fire Brigades Union that there was every chance of an accompanied car on a "shuttle" train catching fire, despite Mr Mitchell's insistence that the Swiss had not experienced a single fire in 34 years of carrying cars with their passengers by train under the Alps [84]. When put to the vote on 13 January, the amendment to require segregation was defeated by the narrowest of margins — nine votes to eight, with all the Labour members (except Miss

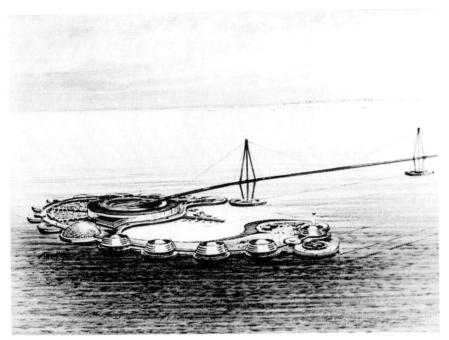

EuroRoute's visionary scheme of bridges, immersed tubes, and artificial islands remained a favourite with French politicians for a long time.

Boothroyd who would exercise only a casting vote) being joined by Mr Aitken and the Liberal Mr Stephen Ross.

The other set-piece in the Standing Committee concerned Waterloo, with Mr Raynsford pressing the case for a public inquiry and Labour campaigning in the country for "dispersal" of Tunnel traffic away from London. This time the Government was on the offensive, following an earlier walk-out from talks with British Rail by Lambeth councillors and local community groups. Councillor Colenutt was accused by Ministers of having "shanghaied" Lambeth representatives into walking out for political reasons, and jeopardising his claims to be seeking a solution to the problem [85]. Mr Mitchell said of Lambeth council that "to go up and down the country complaining about lack of consultation and then to refuse constantly to consult is outrageous behaviour" [86]. Mr Snape added that Labour's front bench found Lambeth's arguments unconvincing, and that he had told the council as much.

The one major change agreed for the Bill during its Committee Stage was the adoption of Shepway Council's alternative access route for the Cheriton terminal, following consultations between that authority, the

Department of Transport and Eurotunnel; the Bill was not actually to be amended until its Committee Stage in the Lords by which time the route had been further modified. Throughout the entire proceedings the Government suffered just one defeat — on an amendment requiring Eurotunnel to provide for the carriage of cycles through the Tunnel. It also accepted an amendment from Mr Raynsford that work on the depot for Tunnel rolling stock at North Pole Junction, beside Wormwood Scrubs in West London, should not begin until local councils and the London Wildlife Trust had been consulted on its impact on a small wood said to be teeming with wildlife [87].

Having been "reported back" to the Commons by the Standing Committee, the Bill now went before the full House for its Report Stage. This usually consists of set-piece debates on the main issues tackled in committee, with the Government bringing forward further tidying-up amendments of its own. Opponents of the scheme had some hopes of forcing changes on the fire safety issue at least, but were disappointed. Although they kept the House sitting well past midnight, their last-ditch efforts to require segregation of shuttle passengers from their cars and force a public inquiry on the Waterloo terminal were not put to the vote. The only amendment to be voted on, a Labour one calling for greater safeguards for the "Heritage Coast" near Dover, was defeated by 151 to 90, a majority of 61. And in the small hours of 4 February, the Bill completed its progress through the Commons after almost exactly nine months when a thinly-attended House gave it a Third Reading by 94 votes to 22, a majority of 72.

The Bill now passed to the House of Lords, which prides itself as a "revising chamber" drawing on expert advice from its members, both hereditary and nominated for life. Once it was clear the Bill was coming, peers were on the receiving end of a barrage of mail opposing the scheme from ferry operators, maritime trade unions and anti-Tunnel groups. And the Second Reading debate on 16 February coincided with a boardroom upheaval at Eurotunnel, with Sir Nigel Broackes resigning and Lord Pennock confirming that he would stand down as chairman once a successor was found (it was to be Mr Alastair Morton). There had also been Press reports, encouraged by Flexilink, that senior Ministers were starting to distance themselves from the project, so that the Government would not get the blame if the consortium finally failed to "get its act together". And when Lord Brabazon of Tara, a junior Transport Minister, moved the Second Reading, Viscount Whitelaw, Leader of the Lords and Mrs Thatcher's unofficial deputy, made a point of sitting alongside him. Lord Brabazon made it clear that the Government was fully behind the project, and Lord Pennock, despite his own imminent departure from the chair of Eurotunnel, made a forceful declaration of confidence in the success of the project despite its problems. And at the close of a debate which did much to restore the morale of Tunnel supporters, peers gave the Bill an unopposed Second Reading.

The stage was now set for a fresh round of Select Committee hearings, with objectors to the Bill as amended having the right to petition the Lords against it. There were four main differences from the procedure adopted in the Commons: the peers on the committee had to have no possible local or business connection with the scheme; they could, and did, decide to hear a broader sweep of evidence than the Commons committee; there would be no need to tackle a number of issues now settled to the objectors' satisfaction; and they did not take evidence in Kent although they visited contentious sites on 11 March. A committee of seven was nominated under the Independent peer Lord Ampthill, a deputy Speaker of the Lords and a director of United Newspapers, which ironically owned the Aitkens' former newspaper the *Daily Express*. Sitting with him were three Conservative peers, Lord Elliott of Morpeth, Lord Gray of Contin and Lord Holderness, two Labour, Lords Galpern and Shepherd, and the SDP peer Lord Harris of Greenwich. Each apart from the chairman was a former Minister, whip or deputy Speaker of the Commons; Lord Shepherd, who admitted to having a cottage near Ashford, was a former chairman of the National Bus Company.

The hearings opened amid controversy over whether enough had been done to remind objectors of their right to petition the Lords. The Department of Transport, going by the book, took the view that having gone through the Commons' procedure and knowing the Lords were to examine the Bill, anyone with an objection would make sure they lodged a petition. But in response to queries from the public, the Department placed reminders in the Press five days before the 18 February deadline. The effect was threefold — to prompt a sudden rush of petitions which up to then had only trickled in, to leave a number of objectors complaining that they had not been able to lodge theirs in time, and to create the impression that the reminders ought to have been issued much earlier.

When the Select Committee met for the first time on 2 March, Lord Ampthill rebuked the Department for its handling of the matter, which he termed a "slight on the petitioners." However, he ruled that there was no way the committee could consider petitions over and above the 1459 submitted before the deadline... roughly one-third of the number submitted to the Commons, although a number commissioned by Sealink had up to 75 signatures each. Mr Fitzgerald, having apologised, went on to rile the committee further by indicating that the Government intended to challenge the "locus" of far more objectors than it had in the Commons, having come in for criticism for letting petitioners get to the witness stand only to be told their evidence was out of order. He listed a series of subjects on which the Government intended to challenge any objector — rabies, terrorism, nuclear waste, Britain's "island status" and the powers of French police in Britain — and others, notably safety, the decline of the merchant fleet and the viability of the project, where it would only accept evidence from those directly involved [88]. It later transpired that a number of petitioners had been sent letters in legal jargon querying their right to appear, which deterred several from coming. Lord Ampthill was not best pleased by the Government's line, and made it very clear to Mr Fitzgerald that the committee would decide what evidence it heard; it was to arrive at a very different view. Such rulings heartened campaigners against the scheme, already encouraged by Eurotunnel's boardroom difficulties and a MORI poll for Flexilink showing only 20 per cent support for the project in Conservative-held marginal seats. But by now they saw little prospect of halting it by political means.

The Lords Select Committee was quick to show its independence. The Government had pressed it to sit five days a week and complete its work by 9 April; the committee resolved instead to hear evidence four days a week in the expectation of finishing by mid-May. Hastening because of the imminent election it was to hear the last witnesses on 30 April, its 30th day of taking evidence. The mood of the hearings was calmer than in the Commons, possibly because the average age of the panel was 15 years or so higher. There was more empathy between Chair and objectors, with many critics being allowed more latitude than when they had addressed the Commons committee and Mr Prosser earning unstinted praise. Even Lord Ampthill, however, had to emulate Mr Fletcher and cut Mrs Litten off in full flight [89]. The chairman also suppressed efforts by a group of mariners from Dover to speak against the Tunnel as such. The less adversarial mood owed much to the chair's readiness to accept evidence ruled out by the Commons committee, for example over rabies. Having heard the expert who explained the precautions to be taken by Eurotunnel, Lord Ampthill reassured subsequent witnesses that their fears were groundless. The committee was also reassured to be told by the Customs and Excise that it had dropped its objections to on-train checks, if only for passengers travelling on through-services beyond London [90]. The lighter touch was reflected in a petition from pupils of Park Farm County Junior School in Folkestone voicing concern about noise, dust and harm to wildlife during construction. And while Mr Aitken, who having been unable to appear before a Commons' committee was free to petition the Lords, responded courteously when told that 90 per cent of the matters he wished to raise were out of order, he still managed to accuse Eurotunnel of "railroading, cutting corners and intimidation", phrases he later withdrew [91].

The Lords Select Committee hearings had completed their first week when on 6 March the *Herald of Free Enterprise* went down off Zeebrugge, with the loss of almost 200 lives. When the committee next met, Lord Ampthill called for a minute's silence, but it was some time before the impact of the disaster affected the proceedings. At first the cause of the disaster was not known, the death toll was heavily underestimated and it took a little while to ascertain that drivers were still in their cabs and that some lorries had been carrying chemicals barred from passenger ferries. The atmosphere when seamen's representatives from Dover testified a few days later was heavy with emotion, but it was before anyone sought to draw lessons for the Tunnel from the disaster. On 1 April, Dr Philip Goodwin of Oxford University's Transport Studies Unit, authors of an early sceptical report on the tunnel, told the committee: "My judgment is that the fate of the Herald of Free Enterprise and its crew and passengers will raise the general standard of safety and attention to detailed procedures, but will very substantially alter the relative economics of ferries and Tunnel" [92]. The committee went ahead with all thoroughness, however, in questioning the fire risk in the Tunnel posed by passengers staying with their cars. And while Mr Drinkwater was to complain of the "virulent campaign" being waged by ferry operators to prevent Eurotunnel raising its share capital, there was a sympathetic hearing for the ferries' argument that they should receive some form of compensation once the link was operating.

Once again there was a battle-royal over Waterloo, this time with Lord Galpern, a canny 84-year-old Glaswegian, taking the part of Mr Raynsford. In advance of the hearing, the London Strategic Policy Unit of Labour-controlled boroughs had caused some amusement by saying in its written evidence that Waterloo was unsuitable as a terminal because British Rail had made no arrangements there for "black and ethnic minority people, gay men and lesbians." A BR spokesman spent some time trying to ascertain what the facilities might consist of, but was unable to [93]. This time British Rail gave full evidence on more central issues, including a demolition of the case for using the Snow Hill link for Chunnel trains which completely satisfied the chairman [94]. Government counsel made much of Lambeth's continuing refusal to meet Ministers or British Rail to discuss Waterloo, at the same time saying they had not been consulted. Lambeth, for its part, produced British Rail correspondence from 1980 saying that Waterloo was completely unsuited as a terminus for Tunnel trains [95]. By the close of the debate, community activists were again claiming that the committee had prejudged the issue.

While the two Houses of Parliament were grinding through their consideration of the Bill authorising the Tunnel project and ratifying the treaty, the French political process had also been under way, though with a lower profile. French procedures were in any case more straightforward, and the task on that side of the Channel was all the simpler because land required for the Fréthun terminal had been acquired by the State prior to the previous scheme in 1974, and never relinquished [96]. Public consultation between the Senate, in particular, and local interest groups got under way at much the same time as the Commons Select Committee, though on a rather less formal basis; indeed M. Henri Ravisse, President of Calais Chamber of Commerce, was able to tell the MPs in July 1986 that he and colleagues had already met members of the Senate's commission, and hoped before long to see the equivalent panel of the Chamber of Deputies [97]. In the first instance the project could be vested with official status by a Declaration of Public Utility, pronounced as a decree after consideration by the President's Council of State. When legislation was promoted at the turn of 1986–87, it took the form of two Bills, one from the Quai d'Orsay ratifying the treaty and the other from the Ministère de l'Urbanisme, du Logement et des Transports formally awarding the contract to the concessionaires and spelling out details of the scheme. Both these Bills were put to the Council of Ministers for approval on 14 January 1987, and in mid-March the commissions of the Senate and Chamber of Deputies, which had been studying the implications of the project, met to compare notes. There was continued sniping by interests which would have preferred EuroRoute, but it made little impression. The National Assembly convened for a new session on 2 April, and the two Bills were then formally introduced, being scheduled to go first to the Chamber of Deputies and then to the Senate; in the event of disagreements — something the previous meeting of commissions was specially designed to avert — the contentious items would be referred to a joint conference of the two Houses. The Chamber of Deputies passed the measure on 23 April 1987 and the Senate on 3 June.

As the Lords Select Committee ploughed through the evidence put to it, election fever was growing steadily and the prospect of a June election loomed. Although the House of Lords is not elected, its work would automatically be halted by the dissolution of Parliament by the Queen so that an election for the Commons could be held. Lord Ampthill and his colleagues thus reported speedily that while 109 amendments were needed to the Bill, none were required that would alter its basics. With the election out of the way — assuming a Government opposed to the procedure had not taken power — the Lords could then carry on and move into the Standing Committee stage, which in the Upper House usually involves the whole House. After Report Stage and Third Reading, amendments approved by the Lords would then go back to the Commons for further, brief, debate, with the way then clear for Royal Assent ... something which is never withheld.

Short of a Labour Victory in the election on 11 June the Bill would probably complete those remaining stages just in time for the Royal Assent at the close of July. Had Eurotunnel stuck to its original timetable for raising £750 million at its "Equity 3" flotation, such a timetable would have raised serious difficulties for the consortium; its decision to delay the share issue until the autumn, taken early in April 1987, removed the pressure. At that point there was still a chance of the Royal Assent by the end of June if Mrs Thatcher opted for an autumn election, but the promoters were not prepared to take the risk. And when public disagreements surfaced between Eurotunnel, British Rail and the SNCF over the terms on which through-trains would use the Tunnel, the extra time allocated was to prove of considerable value. Mr Moore, the sixth British Transport Secretary to have oversight of the project since the "mousehole" was promoted, was looking forward to an early start to construction, provided the financial world played its part. The politicians, save for last-minute armtwisting of railways and financial institutions to make sure the project did not fall at the final hurdle, had done their bit.

References

[1] *Lords Hansard*, 16 Feb 1987, col 923.
[2] *Hansard*, 19 Mar 1980, col 389.
[3] Second Report of the Transport Committee, Session 1980/81.
[4] *The Times*, 8 Jul 1981.
[5] *Guardian*, 12 Sep 1981.
[6] *L'Empire Moderne*, Barbadel & Menanteau, Ramsay 1987] p 183
[7] *The Times*, 20 Jun 1985.
[8] *Hansard*, 21 Oct 1981, col 299.
[9] *The Times*, 10 Apl 1982.
[10] *Daily Telegraph*, 7 Apl 1982.
[11] *The Times*, 17 Jun 1982.
[12] *The Times*, 25 Jan 1984.
[13] *Hansard*, 22 May 1984, col 380.
[14] *Channels and Tunnels*, Sir Nicholas Henderson, p.7.
[15] *Daily Telegraph*, 30 Nov 1984.
[16] *L'Empire Moderne*, A. Barbanel/J. Menanteau p.175.
[17] *Hansard*, 2 Apl 1985, col 1079.
[18] *Commons Standing Committee Hansard*, 2 Dec 1986, col. 13.

[19] *Guardian*, 16 Jan 1985.
[20] *Guardian*, 1 May 1985.
[21] *Channels and Tunnels*, p.41.
[22] *Daily Telegraph*, 15 Aug 1985.
[23] *Daily Telegraph*, 1 Feb 1986.
[24] *The Economist*, 13 Sep 1985.
[25] Commons Select Committee Minutes of Evidence, p.1042.
[26] *Construction News*, 29 Aug 1985.
[27] *Daily Telegraph*, 29 Sep 1986.
[28] Commons Minutes of Evidence, p.502.
[29] *Channels and Tunnels*, p.40.
[30] *Ibid*, p.45.
[31] *Lloyds List*, 6 Nov 1985.
[32] *Daily Telegraph*, 6 Dec 1985.
[33] *Sunday Telegraph*, 8 Dec 1985.
[34] *The Guardian*, 6 Dec 1985.
[35] *Hansard*, 9 Dec 1985, col. 641.
[36] *Hansard*, 10 Dec 1985, col. 760.
[37] *The Observer*, 15 Dec 1985.
[38] *Daily Telegraph*, 29 Dec 1985.
[39] *Channels and Tunnels*, pp. 48–9.
[40] *L'Empire Moderne*, p. 176.
[41] *Daily Telegraph*, 4 Jan 1986.
[42] *L'Empire Moderne*, p. 186–7.
[43] *Channels and Tunnels*, pp. 51.
[44] *L'Empire Moderne*, p. 183.
[45] *Channels and Tunnels*, p. 62.
[46] *L'Empire Moderne*, pp. 189–90.
[47] *The Times*, 18 Jan 1986.
[48] *Channels and Tunnels*, pp. 60–1.
[49] The Channel Fixed Link, Cmnd 9735, p.4.
[50] *Guardian*, 13 Feb 1986.
[51] *Daily Telegraph*, 22 Jan 1986.
[52] *The Observer*, 13 Apl 1986.
[53] *Daily Telegraph*, 7 May 1986.
[54] *Daily Telegraph*, 20 June 1986.
[55] Commons Minutes of Evidence, p.1.
[56] *Daily Telegraph*, 3 Jul 1986.
[57] Commons Minutes of Evidence, p. 139.
[58] *Ibid*, p. 410.
[59] *Daily Telegraph*, 18 Sep 1986.
[60] Commons Minutes of Evidence, p. 983.
[61] *Ibid*, p. 1000–1.

[62] *Ibid*, p. 962.
[63] *Ibid*, p. 1143.
[64] *Ibid*, p. 683.
[65] *Ibid*, p. 1249.
[66] *Ibid*, p. 755.
[67] *Ibid*, p 775.
[68] *Daily Telegraph*, 18 Sep 1986.
[69] Commons Minutes of Evidence, p. 525.
[70] *Ibid*, p. 1143.
[71] *Hansard*, 28 Oct 1986, col. 158.
[72] *L'Empire Moderne*, p. 249.
[73] Commons Minutes of Evidence, p. 1747.
[74] *Ibid*, p. 1674.
[75] *Ibid*, p. 1835.
[76] Commons Standing Committee Hansard, p. 4 Dec 1986, col. 59.
[77] Commons Minutes of Evidence, p. 910–4.
[78] *Daily Telegraph*, 6 Nov 1986.
[79] *Daily Telegraph*, 29 Nov 1986.
[80] *Commons Standing Committee Hansard*, 16 Dec 1986, col. 217.
[81] *Ibid*, 4 Dec 1986, col. 41.
[82] La Voix du Nord, 8 Dec 1986.
[83] *Commons Standing Committee Hansard*, 13 Jan 1987, col. 373.
[84] *Ibid*, col. 392.
[85] *Ibid*, 16 Dec 1986, col. 257.
[86] *Ibid*, 20 Jan 1987, col. 711.
[87]· *Ibid*, col. 716.
[88] *Daily Telegraph*, 3 Mar 1987.
[89] Lords Minutes of Evidence, 10 Mar 1987, p. 21.
[90] *Ibid*, 9 Apl 1987, p. 18.
[91] *Ibid*, 16 Mar 1987, p. 14.
[92] *Ibid*, 1 Apl 1987, p. 7.
[93] *Daily Telegraph*, 19 Mar 1987.
[94] Lords Minutes of Evidence, 30 Mar 1987, p. 31.
[95] *Ibid*, 23 Mar 1987, p. 29.
[96] *L'Empire Moderne*, p. 171.
[97] Commons Minutes of Evidence, pp. 658–9.

Parliamentary handling of the Channel Tunnel Bill

Commons	Lords
STANDING ORDERS' COMMITTEE 6–21 May 1986	
SECOND READING 3 Jun 1986 passed 283–87	
SELECT COMMITTEE 19 Jun–23 Nov 1986	
STANDING COMMITTEE 2 Dec 1986–22 Jan 1987	
REPORT STAGE/THIRD READING 3–4 Feb 1987 passed 94–22	
	SECOND READING 16 Feb 1987 unopposed
	SELECT COMMITTEE 2 Mar– 30 Apr 1987
	STANDING COMMITTEE
	REPORT STAGE/THIRD READING
CONSIDERATION OF LORDS' AMENDMENTS	
	ROYAL ASSENT

3

Tunnel Technology

Bronwen Jones

Holes in the ground are fascinating things. Whether just a cursory glance or a curious stare, we have all at some time watched the road being dug up. Passers-by may be relieved to find good brown soil beneath the tarmacadam veneer of city living yet surprised at the complexity of the urban entrails of sewers, cables and ducting within it. The Channel Tunnel is just a very big hole in the ground, but it will provide a lot more to interest the curious.

Monsieur Jean-Loup Dherse, Chief Executive of Eurotunnel, told a group of eminent politicians, financiers, Government officials, engineers and scientists at a Royal Society gathering in mid-April 1987, that he compared the Tunnel to a bottle of Chateaubriand. "Everyone was interested in what came out of it, but only the glassmaker cared how the bottle was made". But he was proved wrong. Immediately following his talk, came the Technical Director of Eurotunnel, Mr Colin Kirkland, with a comprehensive appraisal of the mechanics of digging, lining and commissioning a tunnel — his audience was intrigued. So the glass is interesting in itself and because it facilitates storage of wine!

There are to be three parallel tunnels stretching nearly 50 kilometres between Britain and France, and many smaller tunnels linking these three submarine bores together. The concept appears simple and very similar to ideas initially proposed over a century ago, but tunnelling is a highly skilled branch of civil engineering and this project will test such skills to the full.

It is not sufficient to design the ideal tunnel from engineering theorems because a host of compromises have to be negotiated to satisfy health, safety, aesthetic and environmental concerns — whether the engineer thinks they are justified or not. Political and legal restrictions have tempered the Tunnel's dimensions, yet technological advance since the last attempt in 1974 has more than balanced the scales. New machines, especially for the French side and improved site investigation, allow selection of the most direct route despite some geologically difficult ground.

Route planning is where the tunnel engineer's work starts, in collaboration with transport and mechanical engineers. The narrowest submarine crossing possible is chosen within the confines of the type of traffic it will carry and the destination of that traffic. Trains cannot travel effectively on steep gradients or rapidly on curves. The stations they serve cannot be moved for the convenience of a tunnel, so the route deviates to satisfy the end user. Gradients permissible are a minimum of 0.18 per cent so that water drains away efficiently, up to a maximum of about 1.2 per cent for trains to work effectively. The change in slope, called a vertical curve, must not be more than of 15 kilometre radius and horizontal curves are restricted to 420 metres radius.

Fortunately, most of the geology beneath the Channel can be, and has been described as, an ideal tunnelling medium. This is not to say that accidents will not occur, but it should be a far easier task than its closest equivalent, the Seikan tunnel in Japan. In March 1985 the Seikan tunnel holed-through, as men drilling and blasting the rock worked from the islands of Honshu and Hokkaido and met in the middle of the Tsugaru Strait. At 53.85 kilometres long (with 23.3 kilometres beneath the Strait), it is one of the longest tunnels in the world and certainly the longest submarine rail tunnel in existence. A 20 year construction period through complex and earthquake-ridden geology, with the loss of 34 tunnellers' lives and a huge accumulated financial debt that may never be written off, proves the technology is not in doubt but the viability of schemes should be very carefully analysed.

In total, 6,330,000 cubic metres of rock were excavated from the Seikan tunnel and 1,510,000 cubic metres of concrete were placed within the hole. The main tunnel was excavated at a gradient of 1.2 per cent from both shores, decreasing to 0.3 per cent in the central section. Six contractors worked in two different joint ventures and used in excess of 12 million man-days of work. After four cave-ins and other calamities, the tunnellers learnt the dificult process of probing over 2000 metres ahead. Their

Roadheader-type tunnelling machine being used to excavate the access tunnel through Shakespeare Cliff in 1974.

experience will aid Chunnellers. After the Seikan tunnel, anything else looks easy!

One of the most complicated areas for the engineers to consider is what to do with the service tunnel at running tunnel cross-overs... whether to go under, over or off to one side, bearing in mind that the service tunnel too has to operate some form of transport system eventually. There are two cross-over sites planned beneath the Channel, another one between Castle Hill and Sugarloaf Hill, another probably within Castle Hill, and one near the French terminal area.

It is the intention to build 150 cross-passages at 375 metre intervals to join the running tunnels to the service tunnels, these to be used among other things for the evacuation of people from the shuttle in the event of an emergency. The passages will be in the order of 3.3 metres wide and fitted with 1.4 metre wide fire resisting doors. So that people have access to the cross-passages, a 1.2 metre wide walkway will be built on the nearest side of each running tunnel to the service tunnel.

UK SIDE DIMENSIONS:
Running tunnels (submarine) internal diameter 7.6 metres
                         outer diameter 8.32 metres

Running tunnels (subterranean) internal diameter 7.6 metres
                                           outer diameter 8.72 metres
Service tunnel (submarine) internal diameter 4.6 metres
                                           outer diameter 5.34 metres
Service tunnel (subterranean) internal diameter 4.6 metres
                                           outer diameter 5.72 metres
The service tunnel will be centrally placed between the running tunnels,
15 metres away on each side.
Concrete segments: 1.5 metres long; eight segments and a keystone for
each ring of lining.
Cast iron segments: 1.5 metres long; eight segments, bolted without
keystone, thicker beneath land, than beneath the sea.
Cast iron for use at junctions with cross-passages every 375 metres.

The rock beneath the Channel in which tunnelling machines will burrow
is called chalk, with a capital C geologists would insist. This is where the
tunneller's saga really begins. More than 65 million years ago, in a time
known as the Cretaceous Period, a type of limestone made of millions of
skeletons of minuscule sea creatures was formed. Ratios of planktonic
(surface-living) to benthonic (sea floor-living) foraminifera fossils
sampled every metre down in 32 boreholes, form the basis of rock dating
across the Channel.

    In the sub-tropical Cretaceous seas, the water where the Channel now
flows was warm, clear, and fairly shallow. Fish were abundant, corals
and crinoids grew on reefs, heart-shaped sea urchins and bulbous
sponges flourished on the sea bed while ammonites curled through the
water. Even the odd delightfully-named reptile, like the *Acanthopholis
horridus*, swam by and dinosaurs paddled by the shore after munching
through a few angiosperm flowers.

    Initially, the water washing the coasts of "France" and "England" was
a little murky, with clay particles suspended in it. The debris, washed
from the low-lying surrounding land, grew less as time went by.

    Floating on or just below the surface of the water, planktonic algae lay,
soaking up the sun. As some of these protozoa died, more grew and the
spherical corpses of their forebears fell unceasingly onto the sea floor. Just
a thousandth of a millimetre wide, each tiny lacy skeletal fragment, called
a coccolith, once revealed to the human eye is infinitely more beautiful
than the most perfect snowflake. This delicate submarine tomb grew
thicker and thicker, the weight of skeletons above crushing the millions of
coccoliths below, into a competent rock layer. Silica from sponge spines,
called spicules, was compressed until it liquefied, flowing into voids in
the rock and resolidifying to become nodular flints in the 240 metre thick
chalk band. The nearest modern equivalent to chalk formation is a

"Globigerina ooze" in deep Atlantic waters or other oozes off the coast of

As in 1974, it is planned to use roadheaders for short stretches of access tunnel to the main workings. It is also likely that the tunnel through Castle Hill will be excavated in this way.

Hawaii. Globigerina is a kind of foraminifera. The British climate must once have been exceptionally pleasant to have had such inhabitants!

In some areas in Eocene times that followed the Cretaceous, sandstones were formed on top of the chalk, followed by clay–sand mixtures, pebble beds and more clay containing fossilised sharks' teeth and the bones of crocodiles who had flopped down in the mud and died. But generally the climate was growing colder as the next ice age approached, the crocodiles knew there was no point hanging around. Only eleven thousand years ago, hairy mammoths trudged on day trips between England and France, churning up the chalk dust and squelching through the peat. Ice accumulating on the main land mass started to weigh it down, but still, only eight thousand years ago, reindeer wandered south for the season, and had no watery barrier to tackle.

Then the land moved down and depressions in the Channel region were flooded, becoming the waterway we see today. These bob-apple-like movements of vast land masses are part of a process called isostatic readjustment that has continued throughout time. Land around the Kent coast is still bobbing upwards in delayed reaction to the weight of glaciers being removed from the English Midlands and further north. Streams are seen to cut sharp v-shaped valleys, trying to maintain their level as the land moves up on each side. The period in which we are now living is merely an "interglacial" ... another much colder spell is inevitable again. It is likely that the Channel will rise once more as dry land, but it should not endanger the life of the concession for the Tunnel owners!

The Strait of Dover transects the northern limb of a Tertiary age anticline. This gently folds chalk and older rocks into an underground "hill" trending east–west. There are several smaller anticlines under the Strait, one south of Dover resulting in Gault Clay appearing in the middle of the chalk. At Quenocs near Sangatte in France, chalk and the clay that naturally occurs beneath it are both folded sharply.

The shape of things that were

Many minor faults with throws (ground displacement) just within sparker test sensitiviy of one to three metres, were detected in surveys over 20 years ago. It was also found that chalk was affected by weathering to appreciable depths below the sea-bed, making it highly permeable and more fissured and broken *in situ* than would normally be expected. Weathering was due to permafrost effects (that now occur in tundra regions beyond the Arctic Circle) dating from the last ice age and a much lower sea level. Movement of faults deep underground in the Kent coalfield may also have resulted in some tectonic fracturing.

To determine the stratigraphy, which is essentially the history of the chalk, the microfossils were correlated. It was decided that larger fossils could not be used because of their comparative rarity making their chance encounter in a narrow borehole particularly unlikely. Calcimetry, the percentage of calcium carbonate in a rock, was used to locate the boundary between Grey Chalk and Chalk Marl. Occurring at about 80 per cent and more marked on the French side, this was used as an upper limiting factor in routeing a bored tunnel. Mott Hay & Anderson, consultants to TML, said in April 1987 that they no longer considered this to be a sufficiently accurate marker and would prefer to use the properties of materials, such as permeability or unconfined compressive strength, determined from testing. The upper boundary of the Gault Clay has been found to be a distinct seismic reflector, and so is easily identifiable on seismic profiles. It is intended to have a target minimum of five metres between the tunnel invert and the Gault Clay.

Chalk Marl, up to 30 metres thick, contains a maximum of 75 per cent of calcium carbonate, with the average at about 65 per cent. Cyclical deposition of sediments occurs with limestone, rich in sponge fossils, every 30 to 70 centimetres. It is good tunnelling ground, with low permeability, easy excavation and the ability to stand unlined for lengthy periods as shown by the Chunnelling attempts from the 1880s. Any permeability *in situ* is due to fissures. Though a tunnelling hazard, these have proved useful in the Folkestone area for water supplies. Small black phosphatic nodules and finely-divided pyrite occur in some layers but are unlikely to have any engineering consequence although their reaction with consituents of the concrete tunnel lining or grout must be borne in mind. For any given chalk layer, the calcimetry and in proportion, the

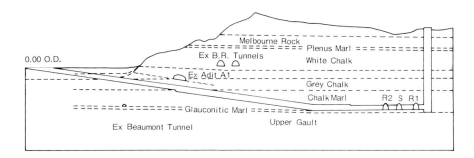

Complicated methods of ground support in short stretches of tunnel will range from the
'New Austrian Tunnelling Method' involving concrete sprayed on the walls, to rock-
bolting and grouting with cement, chemicals, even epoxy resins.

problems, have been found to increase towards France.

Site investigation

The geological detail across the Channel has been ascertained by
sampling, comparison and investigation, almost since interest in a link
began in the mid-eighteenth century. Study of surface outcrops of rock on
both sides of the Channel initially determined the testing areas. Shakespeare Cliff near Dover was found to have a clearly indentifiable
boundary between Lower Chalk and Middle Chalk.

Excluding London Docklands — which would make an Emmental
cheese producer jealous — the narrowest part of the Channel is one of the
best surveyed areas for a civil engineering project in Europe, although on
a much smaller scale, sites for nuclear power stations have more holes
than a pincushion. There are well in excess of 100 boreholes along the
Tunnel route. Holes were drilled for academic interest, correlating
geological records in the UK and France, for electricity cable trenching, for
mineral searches, but most of all, to ascertain the safest, cheapest,
technically feasible path for a Channel tunnel, bridge or some combination of the two.

Early site investigation was perilous, requiring dedication of a level that we no longer seem able to inspire. Pioneers like Henry Marc Brunel, son of the more-famous Isambard Kingdom Brunel, risked life and limb not just in their tunnelling methods but also in pursuit of accurate information on which to base their choice of route. Henry Brunel, was employed by John Hawkshaw to carry out cross-Channel sampling in autumn 1866, following two successful boreholes drilled at St Margaret's Bay near Dover and another about three kilometres west of Calais. Hampered by the bad weather, he worked for six days a week, sampling whenever possible in November and December, and received remarkably little publicity except for a short article in the *Dover Express* which *The Times* copied on 10 November 1866.

When the vessel's crew demanded higher wages for night working, Brunel engaged a new crew. He finished his survey on 24 December, having worked at 207 stations, some for soundings only and some for samples too. The whole survey cost £550 for more than six weeks' work, crew's wages, coal for the tug, insurance, the cost of a gravity corer (the sampling instrument), and sundries. It then took a further month with the aid of sextants and copious notes to plot and record all the results.

Some geologists and engineers pushed themselves to the limits of physical endurance — and when diving technology did not exist, it was simply a case of tying on a heavy bag of stones, jumping overboard and holding one's breath. Monsieur Thome de Gamond claims to have dived to 30 metres depth in this manner in 1855, quickly scooping sands off the ocean floor and shooting back to the surface on an inflated pig's bladder, regardless of getting the bends, before air ran out. In retrospect, the information he gained was too inaccurate to justify the effort involved. There were primitive diving bells, but the full dangers of bends, or indeed similar symptons that occur with compressed air working in tunnels, were not understood, and for many were not recognised until it was too late. Even the men who excavated the last Dartford tunnel across the river Thames near London were not fully aware of the dangers of compressed air until their bones started to crumble away in the crippling disease, bone necrosis. Court cases seeking compensation continued in the 1980s.

Every time the Channel-crossing idea was raised, more site investigation took place. In 1875 over 7000 bottom samples from sites scattered over the study area were collected and examined to prepare the first geological map of the Strait of Dover. Other tunnelling enterprises are detailed in Chapter 7, covering the historical aspects of the Tunnel. In 1957 even the Suez Canal Company became involved. A major bi-national study was carried out in 1964 and 1965, with the Channel Tunnel Study Group (CTSG) — as agents for the French and British governments — reporting

Mini-Eiffel-Tower-like structures were scattered above the Tunnel's route. This one drilled through the chalk and into the clay below, close to Holywell Coombe.

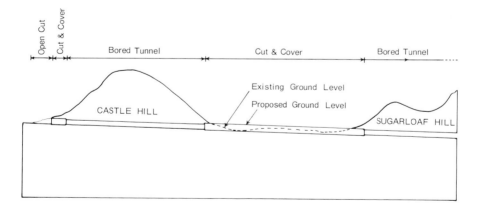

The tunnel across Holywell Coombe, between Castle Hill and Round Hill, will be excavated in cut-and-cover, the trench being carefully backfilled later and the Coombe restored almost to its original condition.

on the feasibility of routes for a bored tunnel in Lower Chalk or for an immersed tube to cross the Channel.

A pilot study by CTSG in 1958–60 with eight successful boreholes had demonstrated that modern developments in marine drilling and geophysical profiling could allow a valid assessment of route feasibility for a bored tunnel in Lower Chalk. But the study took 140 days for the holes to be drilled, of which 77 were lost to bad weather and another 24 days were spent merely fixing buoys required by the shipping authority, Trinity House. CTSG consisted of five consultant groups; Societe d'Etudes Techniques et Economiques (SETEC), Societe Generale d'Exploitation Industrielles (SOGEI), Sir William Halcrow & Partners, Rendel Palmer & Tritton and Livesey & Henderson. CTSG carried out a closely-spaced network of "sparker", "boomer" and "thumper" profiles over 959 kilometres, adding to information gained from 1500 kilometres of geophysical profiling in 1959. Profile trends were north-east to south-west, at intervals of half a kilometre across the study area along twelve tie lines. This network provided a detailed picture of the geology in three dimensions. Further probing by 73 marine boreholes produced about six kilometres of rock core which was recovered and stored in Dover and Calais.

Some of the boreholes were drilled from platforms, the Neptune and the GEM III, while 46 were drilled from vessels. In addition there were nine land bores in France and ten in the UK. Parallel studies of wind, waves, tides, currents, sediment distribution and transport were carried out to assess the feasibility of an immersed tube tunnel and four borehole positions were directly related to this aspect. Immersed tube studies in 1964/5 produced several hundred bottom samples taken at close intervals along the selected route. Siltation of a trench during construction would not be a major problem except in areas of mobile sand waves. "Obviously far more freedom exists to vary the route than in the case of a bored tunnel.." recorded the report of the 1964/5 study. In 1971 and 1972, yet more boreholes were drilled across the Channel. In addition to 17 marine holes, eight were drilled on land, in the Folkestone area. Information on structural geology and the extent of chalk weathering made engineers decide in June 1973 to change the tunnel alignment, not to cross beneath Dover Harbour but instead to pass under the coast at Shakespeare Cliff.

In 1986, Transmanche-Link (TML), the consortium of five French and five British companies with the design and construct contract for the Tunnel, decided to carry out yet more site investigation. Borehole samples had been taken at between 500 metre and 1000 metre intervals along the sea-bed and at roughly 2000 metre intervals on land. TML followed these up with a marine seismic geophysical survey in August 1986 on the proposed alignment and extra boreholes were drilled from

December 1986 until April 1987. Generally the route follows that established for the 1974 scheme, on the UK side.

Geocean, a French company, first tried drilling from a ship but owing to
difficult Channel conditions it changed to use of a large oil platform.
These platforms had become available owing to the downturn in North
Sea oil production and were competitively priced. Once in position they
were unaffected by weather conditions. Geocean chartered a jack-up drill
platform, the Zapata Scotian, to drill down to 60 to 100 metres beneath the
sea floor; the work was expected to take about 38 days to complete six
boreholes (one in UK water, five in French waters) or 60 days if it was
decided that nine holes should be drilled in total. The intended work
programme was for the first hole to be started on Friday 14 November 1986
and filled with cement and sealed once complete. No holes were located
in the main navigation lanes but precautions were still taken against
collisions.

Industry rumours of difficulties related to drilling and other contracts
are difficult to substantiate and may well, rather than being untrue, be
merely a matter of perspective. It was said that both French and British
seamen demanded compensation for disruption of their prime fishing
grounds, which conveniently shifted to wherever the rig was; it is claimed
that French authorities refused the right for British contractors to have
any drilling work in their waters; and TML refutes comments that
Geocean's contract went awry and that part of its work was rewarded to a
UK company.

Land & Marine in collaboration with Safe Offshore was brought in to do
the UK holes required on two sites. In early December 1986, its self-
elevating Launcelot platform sat "ready to mobilise from Holland,
awaiting weather..". A platform has to go down into the water to be
moved to a new location; if a certain wave height is exceeded it cannot be
moved. Land & Marine could not tender for the mid-Channel work as the
Launcelot can only work in water up to 30 to 35 metres deep.

Cores of 85 millimetre diameter were collected. Wire-line techniques
were used to establish a range of engineering parameters and *in situ*
packer tests determined the rock permeability, while geophysical logging
provided more information on geological structure and rock density.
When Transmanche-Link awarded contracts for borehole drilling in
December 1986, it stressed the importance of sealing the holes. These, at
least, are not likely to endanger the tunneller.

Drilling at that time of year was difficult, particularly in waiting for
suitable weather windows when equipment could be moved onto site.
But, most of all, TML and Eurotunnel could not afford to wait; if the
money and the will were available, they had to press on with any

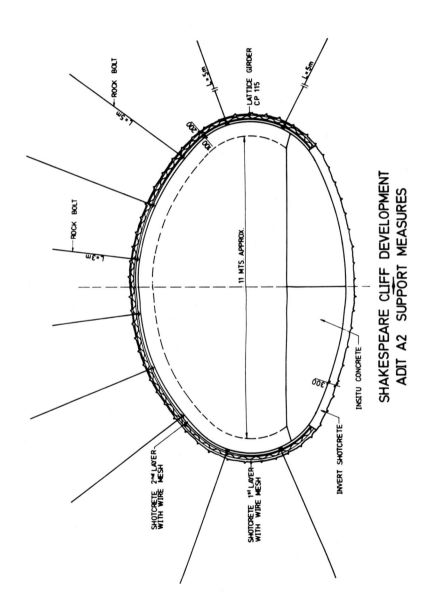

SHAKESPEARE CLIFF DEVELOPMENT
ADIT A2 SUPPORT MEASURES

Pressures of the UK environmental lobby did not allow the French option of a massive 55-metre diameter, 120-metre deep shaft from which to start construction work. Instead they have a spaghetti of smaller tunnels and shafts planned in the base of Shakespeare Cliff.

achievable task or the Tunnel would be pie in the sky rather than fish in a dish. Good political progress in May and June 1987 aided construction planning.

The boreholes were used to correlate the geophysical survey which otherwise would only have had an accuracy within three to six metres, but combined could achieve an accuracy of two to three metres. They were intended to check problem areas on a slight realignment of the Tunnel since 1974 and also to check geology at selected locations such as the 1.5-kilometre-long running tunnel cross-overs, each about six kilometres off shore, in which the trains will change tunnels, the position of pumping stations where water will be gathered by gravity and areas of potential uncertainty as regards the level of the Gault Clay or possible fracturing. If an anomaly was seen on a geophysical profile, a borehole was used to "check it out" and, finally, Mr Gordon Crighton, engineering manager for TML, said "engineering judgement" was used wherever necessary in the choice of a few extra borehole locations. The resolution is superior to the 1973 work, which was the main justification for a new survey. Also said Gordon Crighton "We do not want to stint on the geophysical and geotechnical information".

Drilling technology has become increasingly refined in recent years. By measuring how difficult — or how easy — it is to drill into a certain rock, the driller can get almost instantaneous information from his borehole instead of waiting for the results of laboratory testing. A French company, Samega, has electronic back-up that records rate of drilling penetration, weight on bit, injection pressure of the drilling fluid, rotation speed and torque. This in turn identifies the rock type being drilled, gives characteristics such as hardness and permeability and detects any unusual cavities.

Another contract worth in excess of £100,000 was awarded for drilling in the littoral zone, between low and high water on the UK side. This work was undertaken by Terresearch, a subsidiary of Taylor Woodrow, one of the five contractors in Translink. In the House of Lords Select Committee in April 1987, the possibility of this rig striking wrecks ... particularly those with explosives on board... was raised, not out of concern for drillers, but instead for any passing mariners.

Foreshore drilling took place over ten weeks, along 1.5 kilometres of coastline, between Shakespeare Cliff and Abbot's Cliff near Folkestone. About 14 holes were drilled from a jack-up flexifloat platform 200 metres offshore, using rotary and cable percussion techniques. A further 85 boreholes and 120 trial pits were drilled and excavated on the land sites around Ashford and Folkestone, for information prior to construction of terminal buildings, earthworks and rail or road access. Wimpey also had a contract to drill 29 holes along the tunnel approach line on land.

Considerable on-land drilling and trial pit digging was undertaken for site investigation in the terminal areas on both sides, along the subterranean tunnel route, at proposed portal (tunnel entrance) locations and at the many sites for associated infrastructure, such as new bridges, viaducts, and so on. Kentish fields from January to June 1987 were scattered with mini-Eiffel Tower-like structures and the muddy but temporary scars of tracks used to bring the drilling rigs in. Mr Gordon Crighton stressed how hard the contractors strove to cooperate with environmental bodies. He said "the drilling rigs were placed on wooden planks set on polythene sheets, to protect the grass. All routes to and from the boreholes for machines were agreed with all interested environmental bodies before the rig was moved into position." Sensitised by the effects of prolonged exposure to the political limelight, he vehemently pointed out "no borehole contract has ever been carried out with such care to the fauna...Great care was taken to minimise effects on the landscape."

The best site investigation tool the tunnellers will have is the tunnel itself. Not as foolish as it may sound, the central or service tunnel is just an oversized horizontal borehole, probing into the sub-Channel geology five to six kilometres ahead of the other two tunnels.

Modern submarine geological investigation, after looking at any records already available, starts with a geophysical survey or, more usually, a series of lines are surveyed in this manner. A ship travels above the intended line of the tunnel — or, if this is not decided, it tends to survey the shortest route possible between two land masses. An explosion is detonated in the water and a series of geophones or hydrophones (acting like microphones) pick up the signals emitted. Following similar behaviour to an earthquake shock wave, the pulses from the explosion radiate outwards until they meet an obstacle. This does not stop them, but the difference in density between what the wave was travelling through and the new material or "obstacle" causes a definable deviation. Passing from water to rock is easy to see on the "picture " produced when the signals the geophone detects are plotted on a seismic profile. A series of white, grey and black lines are produced to represent the sea-bed topography and the different layers of rocks below, in a cross-section or a map of the underworld.

Studying this map beneath the Channel shows most of the rocks to be fairly horizontal and that there are few major breaks (faults) or curves (folds) present, although there is a large alluvial depression mid-Channel known as the Fosse Dangeard which the Tunnel has to curve west to avoid. The technique combines geological knowledge and rules of physics, such as those for refraction.

Part of the 1974 workings beneath Shakespeare Cliff. Source: Eurotunnel.

When the seismic profile is available, boreholes are drilled wherever an anomaly occurs, or at regular distances to correlate the information. Seismic profiles identify a change in rock type, but a borehole tells one what kind of rock is actually there by recovering a sample of it. There are also some down-the-hole tests carried out to work out which way the rock is dipping or how waterlogged it is. The results are extrapolated towards the next definite geological record, that is the next borehole, and then compared to the seismic profile. Where results do not tally, a fault whereby a near-vertical slice drops one layer lower than another in a step-like structure, or a fold where the rock has been squeezed into a curve by mountain-building presures, may be present. There are both faults and folds en route, and both could be potential problem areas. The amount of throw (literally the height of the step, to continue the analogy) of a fault is not expected to be more than a metre or so. In heavily-faulted geology, high throw could mean one was tunnelling in chalk and then found oneself trying to tunnel through dolomite. Nothing that dramatic happens beneath the Channel. The problem with a minor fault for the tunneller — or, more likely, a batch of minor faults — is that the rock tends to be weaker in such zones and, rather than a sharply defined plane where one rock has slid past another, there is a zone ranging from say one centimetre to one metre wide of crushed rock and sandy infill. Water can flow along such zones.

It may prove possible to determine the degree of chalk weathering beneath the sea-bed from the degree of distortion of profile records, due to the difference in sonic velocity in weathered and unweathered chalk. Eleven drilling sites were selected across the Channel in December 1986.

Typical profile:
Lower Chalk 80 metres thick
White Bed (White Chalk)
Grey Chalk 40-120 kg/cm squared
Chalk Marl (Blue Chalk)   34-93 kg/cm sq. less than 60% calcite
Chloritic Marl   40-139 kg/cm sq. more than 60% calcite

Experience in chalk

There have not been many tunnels in chalk from which engineers can draw experience. There have obviously been the short stretches of previous Chunnel attempts; Beaumont in 1882, Whitaker in about 1922, a short 19th century adit at St Margaret's Bay — east of Dover —, the UK side access tunnel to the 1974 workings at Shakespeare Cliff (excavated by roadheader, not full face tunnel boring machine (TBM)) and the short Chunnel of 1974, some gas-storage caverns at Killingholme near Humberside on the east coast of England where one of the engineers said "it was not like 'normal' chalk... the rock was as hard as the hobs of hell", a UK Transport and Road Research Laboratory (TRRL) trial Chunnel in Chalk Marl at Chinnor, French Chunnel attempts, shafts through the chalk for access to coal seams much farther down, a 2.3-kilometre-long marine outfall at Brighton in southern England, and perhaps most importantly, the metro project in Lille, in mid-France. About 13 kilometres of the Lille metro are already operational. An extension to the tunnels is expected to be complete in 1988. It was undoubtedly seen as a testing ground for equipment that could be used to excavate at least the French side of the Tunnel. Two different TBMs were used — one manufactured by Fives-Cail Babcock in France for the Japanese company Kawasaki Heavy Industries. The other machine was of West German design and manufactured by Westphalia Lunen. Called a Luchs FL 5R 100 selective cutting machine, it excavated the wet chalk with flint inclusions, in 300 millimetre diameter lumps, and achieved an advance rate of 94 metres a month, working two eight hour shifts, five days a week.

If trucks had been used for chalk spoil, they would have been bogged down in their own tracks as they transported their wet, sticky load. So a new approach to the spoil disposal was selected for contractors SGE and

Cast iron tunnel lining segments being loaded to breaking point. Intensive testing determines the factor of safety required for each component of the Tunnel system.

Fougerolle by specialist sub-contractor Delta Pompage. A method of pumping chalk away as a slurry was adopted and has proved highly successful. Using West German company Putzmeister's equipment, the excavated chalk is taken in large lumps on a short conveyor to a mixer trough and crusher where it is broken down and mixed with water. The thick paste is pumped, sometimes like toothpaste, sometimes like runny porridge, 600 metres along the tunnel in a 140 millimetre diameter high pressure delivery line and then 30 metres vertically to a silo at the top of an access shaft. Aggregate of up to 80 millimetres diameter can be handled and pumped at 60 cubic metres an hour.

Lorries drive beneath the silo and, once full, take the chalk to the disposal site outside the town. Sometimes the hewn chalk is in particularly large lumps which have to be broken manually with a pick, before being turned into slurry. To have spoil removed along the tunnel in a relatively narrow pipeline improves the working environment and safety, there being no need for muck wagons and the accompanying clamour and dust, and it leaves much more space within the tunnel.

The other industry that has had considerable experience with chalk is the North Sea oil and gas drilling industry. They too have to know how the constituents of chalk are formed, how they are accumulated and sometimes resedimented, about dewatering of chalk, its intrinsic and later diagenesis, and how these factors relate to its porosity and permeability. Ground collapses beneath oil platforms, like the continually-subsiding Ekofisk, have been associated with chalk reservoirs.

Full face boring machines and their route between France and the UK

Diving through three main rock layers, Upper, Middle and Lower Chalk, the tunnellers will find the upper one, being the youngest to be the most pure and problematical. Lower layers have a higher clay content, almost up to 50 per cent towards the base. This mixture proves easy to tunnel in, relatively soft and impervious to water inflow, but it provides a waste product that is difficult to dispose of. Close to the shore and the subterranean sections will prove the most difficult areas to tunnel, descending through variable loading and cutting across the geological boundaries.

The three main tunnels will be excavated by 11 full face tunnel boring machines (TBMs), five French and six British. A central service or pilot tunnel will be hollowed out first, with the two running tunnels following about five to six kilometres behind. The TBMs will be made in at least eight different diameters, depending on whether they are to work under land or sea, how the tunnel is to be lined and what kind of geology is anticipated. Among tunnellers, machine design is considered one of the more interesting parts of the project, being literally at the forefront or cutting edge of technology. So, also it is one of the more disappointing parts in 1987, especially on the UK side. The late Sir Harold Harding, tunneller extraordinaire said "the longer the tunnel, the more attractive is the use of the right machine". He was full of praise for machines used in 1882, and on the most recent Chunnel attempt.

In 1974 a Priestly full face tunnel boring machine (TBM) was used. Pushed forward by hydraulic rams, it cut the chalk with double-headed tungsten carbide-tipped rocking picks. The first two machines ordered for the UK service tunnels in August 1986 have been manufactured by Glasgow-based company James Howden, but designed by its Grosvenor Tunnelling Division consisting essentially of the people who designed the Priestly machine. The new machine body telescopes as excavation proceeds, but not much else has changed.

One interesting controversy arose over patents for the pick design, still held by Edmund Nuttall — a company that had hoped for part of the Channel Tunnel "action". The clever part of the design is not so much the picks (which fortunately got over teething troubles in which the tungsten carbide used to fall off) but in their sealed rocker housing that holds the shank of the bit and the synthetic rubber seal which prevents chalk from falling into the mechanism and jamming it. Grosvenor approached Nuttall with a formal request for the right to use the patent and will pay a royalty for it. The normal sort of figure would be about five to ten per cent of the cost of an article. Nuttall said "the Priestly name and Priestly technology were used in 1973. We're quite capable of mobilising that again but we would need to cooperate with someone." It is sad that companies which persevered so long in pursuit of the cross-Channel dream end up with such a minor role to play.

Priestly Channel Tunnel machine: specification

It was a shielded full face soft rock machine with an outer diameter of 5.27 metres and cutter head power of 560 kW. The reversible cutter head speed of 4.5 revolutions a minute could be achieved with a maximum thrust of 1600 tonnes. The machine weighed 150 tonnes and the segment erector weighed a further 100 tonnes. The cutting tools were drag picks. In Lower Chalk of up to 14 MN/square metre strength, the penetration rate was six metres an hour and the time taken to build one ring was 20 minutes.

Cross Channel Contractors ordered the machine to drive the Channel service tunnel and to line it with an expanded precast concrete segmental lining, 360 millimetres thick. The compressive strength of the chalk was expected to be in the 7–14 MN/metre squared range and therefore could be cut using drag picks. It was recognised that to achieve the progress rate required, a fairly sophisticated method of segment handling and erection was needed. The machine incorporated a thrust anchor ring to allow excavation to be concurrent with, and independent of, lining erection. The erector was pushed forward by its own jacking system and was not mechanically connected to the machine. Partial pre-erection of some rings considerably speeded up the machine's progress compared to its contemporaries.

It was planned to use the machine for 20 kilometres of tunnel driving but the project was abandoned by the UK Government after only 300 metres of tunnel. Even on this short length, however, the machine demonstrated that it could excavate at the required six metres an hour and handle and erect four-tonne (1.25-metre-long) segments rapidly, reaching a top performance of ten metres an hour.

Additional features included a high capacity main bearing and integral drive gear, double row labyrinth bearing seals and easy access to the face through the cutter head centre. An anti-stall system was used to maintain high production, and a hydraulic power pack mounted within the machine body maximised erection space. One vertical steer shoe was fitted. Sir Harold Harding described it as "this beautifully made machine" and said that it "emphasised the accelerating sophistication" of TBMs. He commented that "long drives under the Channel in almost ideal material would give scope for trying out two or more machines. As driving progressed it would be no great matter to discard a slower machine and replace it with a faster one". These attitudes may not be totally realistic,

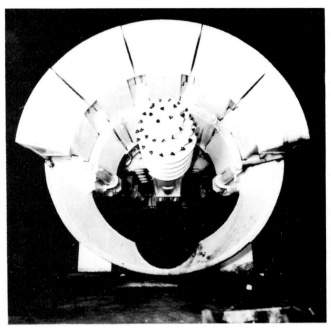

An example of a modern roadheader that could be used on the adit construction.

although certainly idealistic, but Sir Harold raised a major point in his comment that "the philosophy for such long drives cannot apply in the same way to the commoner short drives in other work", and of course, the opposite applies too.

Howden's machine

James Howden of Glasgow and its Grosvenor Tunnelling Division initially had a letter of intent rather than a contract for the first two small TBMs required for the Tunnel. The contractor TML could not make firm purchases until it had sufficient money in the bank after the Equity 2 share placing. The machines will be faster than the Priestly model, designed to line and excavate at a minimum of six metres an hour or to erect one complete ring of lining segments (1.5 metres long) in 13 minutes. Faster speeds than those specified are anticipated. Conveyors to follow behind the machine will be of UK manufacture.

The double-headed tungsten carbide reversible picks in a sealed housing are on four spokes for the service tunnel machine (eight spokes for running tunnel machines) on the cutter head. Seventy picks will be used on smaller machines but if ground conditions change dramatically these could be changed to about 35 roller cutters. One roller cutter could be used to replace two picks from the rear of the cutter head, and at a similar cost to equip the machine. Howden, however, anticipate that their cruciform of 70 picks will not need any changes during its passage across the Channel. So, the picks rock back and forth, dislodging the chalk and sending it though the cutter head onto a retractable conveyor belt (with a loop take-up allowing a 1.5 metre stroke) and thence on to a series of a further three conveyor belts, before dumping it over 100 metres behind, into muck wagons.

The machine specifications were not available at the time of writing but are expected to be similar to many of those for the Priestly machine, the main differences being in the Swiss-design erector and that the machines, as a consequence, will be much quicker. They will also be safer, with almost no unsupported ground and a novel steel-blade ring that cuts off any water flow from the face in times of danger.

Service tunnel machines:　Diameter 5.34 metres "marine" machine
　　　　　　　　　　　　　5.72 metres "land side" machine.
Both about 11 metres long and weighing 350 tonnes.
Running tunnel machines:Diameter of about 8.32 metres (marine)
　　　　　　　　　　　　　8.72 metres (land).
Weights 678 tonnes and 700 tonnes.

The cutter head of the Howden TBM is pushed forward by hydraulic rams and a shield-like tailskin prevents workers being exposed to any unsupported ground. Contiguous with excavation and spoil removal is segment erection. This is achieved by small cranes hooking into inserts and picking up eight heavy segments — delivered in the correct order by wagon — putting them in position and then, without grout or bolts, wedging them into position by means of a tapered keystone. This allows the strength of the surrounding rock to contribute to the strength of the lining and provides a smooth finish that needs little further treatment for the final tunnel.

The segment erection system for the Howden machine is designed by the Swiss company, Robert Waelti. Behind the main boring machine a further 200 metres of vital back-up equipment extends down the Tunnel; conveyors near the ceiling, known as the crown, and supply wagons on rails on the floor or invert. Close to the machine, segment wagons travel within a very large wagon called a rolling deck. Two of these have been built for the UK service tunnel construction by Somerset-based firm, Decon. The chalk-carrying conveyors will be centrally placed, and the one furthermost from the TBM will deposit chalk in a muck wagon, retract to fill the next muck wagon and so on, until one train is full and a California switch system brings the next train into line.

A full face rotary tunnel boring machine fitted with roller cutters, of a similar size to the type that will be used on French service tunnel excavation. This machine is made by Robbins of the USA, the company that won the first machine manufacture order on the French side, in December 1986.

These wagons have been designed by tunnel specialists Muhlhauser of West Germany, but most of them are being built by Howden in Glasgow, in a joint venture agreement. Six locomotives to pull the muck-wagons have been built by Hunslett of Leeds in a £1.2 million contract. To cater for the steep inclines on the 1974 adit - and on another access adit yet to be excavated by a roadheader (frequently used to excavate mine tunnels), these locos employ rack-and-pinion techniques. The same principle is used on some of the tourist trains on mountain routes, and Hunslett supplied the locos that climb Snowdon, the highest peak in Wales.

The tunnel locos are battery- and trolley-powered, with supply from an overhead cable, and they are rated at 350 horsepower. There will be one loco at each end of a train, one pushing and one pulling, and on the access at Shakespeare Cliff, one loco will pull about 100 tonnes on the existing slope with its gradient of 1 in 6. Gradients will be almost flat for the main tunnel, only the access can be so steep. Mr Ken Wainwright, joint managing director of Hunslett, said it took about seven months to manufacture one of the locos as they were "quite a new design". More than 100 locos will be required on the whole construction project.

Longshore drift

It is dangerous to alter the shape of the coastline or to dredge large quantities of aggregate from the sea floor, as the effect on movement of tides and currents combined in longshore drift may have dramatic results many kilometres away. The principal reason for disallowing Tunnel spoil dumping at sea was justified on "protect the fish and seaweed" grounds. However, the effects on longshore drift with several extra million cubic metres of sediment, even if much of it went into solution, could have been severe. Despite even schoolchildren learning of such forces in their geography lessons, mistakes continue to be made. Mention of recent examples is not appropriate here.

The first time that such effects were well documented along the shores of the Channel was in 1917, but the story started in 1896 when it was decided to build a huge new dock at Keyham, near Plymouth. The project required 395,000 cubic metres of broken stone or shingle for the concrete walls. Little suitable material was available on land in the vicinity, apart from rock that would have to undergo an expensive crushing process. But offshore, there was good quality, cheap and easy-to-extract shingle in infinite supply. Contractor Sir John Jackson was granted a licence to dredge, and work started north of the village of Hallsands, in Start Bay.

Dredging took place between low and high water mark, as it was too sandy in the deeper water. The shape and angle of local beaches were altered by the removal of 1600 tonnes of shingle each day, eventually a new low water mark being above the old high water mark. Protesting to their MP, villagers had their fears raised in Parliament but Sir John claimed that natural processes would restore all to normal in time; he won the case.

By late 1900, the protective shingle bank in front of Hallsands had been eroded away and winter storms attacked the rock platform on which the village was built. Remedial measures came too late. Within two years the village was clinging precariously to the edge of the platform. Fishermen had to lift their boats up to the village street because there was no longer any beach on which they could rest. A sea wall was built and minimal compensation payments were made to those whose property had been damaged or lost. A gale in 1917 was almost the end of the story. Four of the remaining 25 houses were demolished almost immediately and 20 others were left uninhabitable; only one house of what was once a self-contained community remained usable. One callous contractor left his mark on the coastline for decades. The small town of Torcross had its foreshore ripped away in 1951 and sustained further major damage in 1979. Seawall

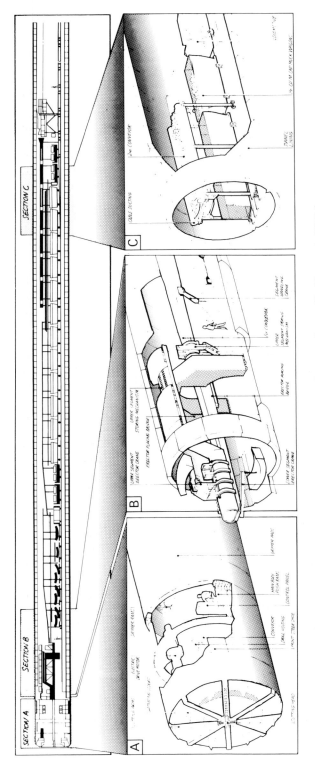

SECTION A

SECTION B

SECTION C

A

CUTTING HEAD

FRONT STEERING SHOE

DRILL HOUSING

CONVEYOR

CONTROL PANEL

MAIN BODY PUSH RAM

COPPER RING

ELECTRIC DRIVE MOTOR

SPINNER PLANET

REAR STEERING SHOE

B

UPPER SEGMENT ERECTOR CRANE

LOWER SEGMENT LABELLOR CRANE

ERECTOR PLACING DEVICE

UPPER SEGMENT STEERING MECHANISM

ERECTOR RACKING DEVICE

LOWER SEGMENT STEERING MECHANISM

NEW CONVEYOR

SEGMENT HANDLING CRANE

C

NEW CONVEYOR

CABLE DUCTING

TUNNEL LINING

NEW CLEAN AND WORK ROOM ???

Schematic diagram of one of the two full face tunnelling machines that James Howden manufactured in Glasgow in 1987, for use on service tunnel excavation from the UK side of the Channel.

technology is now sufficiently advanced to protect this settlement until the next gale.

The reason for telling this tale? Just to forewarn that the effects of the spoil platform protruding from the base of Shakespeare Cliff must be as carefully watched for changes to longshore drift as those that could have occurred if spoil had gone straight into the sea. Any marine aggregate dredging proposals in either France or the UK should be considered with equal care.

Spoil

The UK water authorities fear that salts may be leached from spoil excavated beneath the Channel and deposited on land. Spoil salinity is not yet proven, with only one test carried out by early 1987 on a sample from a hole sunk in the sea-bed. There is argument among geologists as to whether or not any salt content would be leachable. All UK sub-sea spoil will be placed at the foot of Shakespeare Cliff, so the problem will not arise. Conveyors for long-distance spoil-removal in the Tunnel were a possibility but would require a great deal of maintenance. Muck wagons should ideally be dual purpose — segments in, muck out.

The original platform was constructed for Dover Colliery. Its extension is unpopular, but Colin Kirkland of Eurotunnel said "It is not economic in terms of transport costs and requirements for permanent filling to follow up one disposal method mooted, whereby the chalk spoil could be used as disused coal mine infill in Kent".

4.5 million cubic metres in the UK, 3 million cubic metres in France. Extension of the platform at the base of the cliffs for British spoil disposal will eventually be landscaped as an amenity area. On the French side the slurry will have to settle for many years behind a dam. A senior engineer commented "It will be like a valley full of yoghurt; we have no idea how long it will take to solidify ... if it will at all". One West German company has come up with an unusual spoil disposal suggestion, whereby pumps usually used to transport concrete, or spoil in slurry form, along the tunnel, should instead transport the spoil vertically in boreholes, towards the sea-bed. If it is not environmentally acceptable to dispose of the spoil on the sea-bed (and despite strong pump pressure, it does leave a fear of a direct path between sea and tunnel), then the company suggested, that a pocket be created by the pressure of the slurry, anywhere between 10 to 50 metres beneath the sea-bed. It is suggested that the sea-bed will rise in a hummock over the spoil pocket, but unlike the subsidence and heave problems associated with some mining methods, short of the fish getting together a petition, there will be no one to object. Each disposal point would have to be between 500 and 2000 metres apart, depending upon the geological conditions. The 200 millimetre diameter hole could be

Drilling platform when it first moved on-site off the coast from Shakespeare Cliff. It was used to determine the best location for a seawall to retain about 3.75 million cubic metres of chalk spoil.

flanged directly to the TBM and its delivery holes would be lined with steel pipes, and would be between 20 and 40 metres long. This immediately calls for quite a thick overburden above the tunnel and further development of specialist drilling technology used in the oil industry. The company's ideas were not taken up, despite its statement that "the Transmanche pilot tunnel will give a chance to demonstrate this revolutionary idea to develop a basis for huge savings when building in the main tunnels".

Right on target

Modern surveying, especially underground, uses electronic distance measurement and lasers extensively. Computers too are playing an increasingly important part. The laser system used in the Channel Tunnel can really only be supplied by one company, for two reasons. Zed Instruments is the only company to have taken such a dedicated and detailed approach to the task and, the way the tender for machines was written, the designers had no choice but to opt for the only system that complied with the detailed specification. But, it is a good system — so no one can really object. Mr Keith Valentine of Grosvenor Tunnelling worked closely with Zed on the Tunnel project, to ensure at least the first two service tunnel machines will reach their destinations.

A target and the laser emitting the beam will be located in the tunnel crown, but offset from the centre so that any chalk dust rising from the conveyor does not weaken the beam's intensity. The laser guidance system does not just keep the machine travelling forwards in the right direction, but watches for machine pitch and roll. Any tiny deviation extended over 50 kilometres could have drastic effects. Total permissible deviation (construction tolerance) for the tunnel is 150 millimetres. Anything more and the French tunnel machine might not meet the British one, the gradient could become too steep for trains or a curve in the route be too sharp. Apart from the fact that dramatic curves, horizontal or vertical, would slow future trains down, in the short term they could also require special concrete segments and moulds to be manufactured. The consequences of going off-line are too complicated to contemplate, so laser guidance is essential!

There are strict regulations regarding the intensity of laser beams used in mine surveying; surprisingly they are not the same as for civil engineering tunnelling, where in fact the safe limit is far lower. A laser, usually of one to two micro watts helium–neon mix, is fixed to the tunnel wall and illuminates a target unit attached to the TBM. Two inclinometers on the TBM measure roll and pitch with information from the target unit and inclinometers processed so the machine operator can see a display showing his deviation from the planned tunnel axis. He can also see the roll, lead and look-up, which is the angular deviation from three theoretical axes and his predicted position (where the machine will be in five metres if no corrections are made). Using this display the TBM is steered back on course.

Zed Instruments has designed two advanced systems, the TG 260 and the TG 261, one of which will be used in the Channel Tunnel. The TG 260 system has two transducer units fitted on to the machine, a target unit and an inclinometer unit. The target unit receives a beam from the laser and measures the position of the laser spot on its screen and the angle of beam incidence which will tend to the right or to the left. The inclinometer unit measures the roll and pitch of the machine. A third transducer unit underground measures the distance up to 600 times of the machine along the designed tunnel axis (DTA).

If basic criteria are not met, they are shown in reverse video until a correction is made. Information from all transducers is sent to the below-ground Engineer's Unit for processing, after which it is transmitted to the above-ground computer (ACU) by direct cable link. Position and attitude of the machine are calculated and results are displayed on the visual

Early days. Construction just under way on the French shaft from which six tunnel headings will be driven, three landwards and three seawards. This enormous hole was classified as 'temporary' works.

display unit above-ground as well as being displayed underground with measurements in millimetres, and symbols identify the property determined as well as in which direction it is tending, if at all.

Horizontal displacement, vertical displacement and lead are measured relative to the laser beam; roll and look-up are relative to the vertical direction; axial displacement is relative to an arbitrary starting point. Each time the laser is re-positioned, new global co-ordinate values are dialled into the below-ground engineers unit. Data is stored even when power is switched off and information can be printed out on command or at automatically set intervals.

Self-diagnosis of problems identifies misalignment of the laser or transducers, or cable severage. Flashing numbers indicate that data may no longer be valid. Cables for the laser guidance system are a conundrum. For a direct link to the surface, the cables will cost as much again as the whole system and are also the most likely cause of a malfunction through interruption of the signal or other disturbance. It would be possible to transport the information on a computer disc on one of the underground trains, if engineers did not want instantaneous information.

"Robust" or "rugged" are overworked claims for high technology used in any environment other than a pristine hospital operating theatre. But for use in a tunnel, the equipment really has to be rugged, and it is difficult to combine such qualities with delicate electronics that are highly sensitive to the ever-present moisture, dust and temperature variations. Zed Instruments even commented that on a specification for British Coal it was asked to make a target that would, say, not be damaged if a man wearing a diamond ring chose to etch his initials on it. Vandalism by the user knows no bounds, but it is difficult to design for every eventuality. The company won a Queen's award for technological achievement in 1986. Perhaps as its lasers beam into French territory, it will also win an award for exports!

Route

Apart from an underground railway loop within the terminal area at Cheriton, in the UK the Tunnel will first go underground at Castle Hill, on the edge of Folkestone. After burrowing through the hill it will cross a controversial vale called Holywell Coombe in cut-and-cover. This is a tunnelling method whereby a large trench is dug, the tunnel tube formed in concrete and the trench filled in again, sometimes leaving an embankment above.

In allowing this part of the text to be checked for technical accuracy, I encountered the paranoia that grips almost all engineers involved with the Tunnel. While they are concertedly trying to design a project that meets demanding engineering standards, they are painted as ogres seeking to despoil the countryside. So in case some trigger-happy journalist with no technical understanding takes a word like "sometimes" in discussion of a general principle and carelessly leaves it out in a front page story, the engineers would rather I did not explain any general procedure or discuss any "what if" scenarios. Unfortunately I cannot be bound by such requests, merely strive for as accurate a portrayal of fact as possible.

To return to the Coombe, the backfill will be carefully reinstated and the area landscaped and planted with appropriate vegetation. Mr Gordon Crighton of TML said "From the surface there will be no evidence that the

Tunnel passes underneath whatsoever!" Landscaping of Holywell Coombe is a contractual obligation of TML and great attention was paid to the final scheme at the design stage. All the relevant environmental bodies and planning authorities were consulted. Putting the Tunnel in

High technology laser guidance systems will ensure the three tunnels do not stray more than a maximum of 150 millimetres off course.

cut-and-cover across the Coombe allows the correct railway gradient to be achieved along the route, although lowering of a viaduct beyond the terminal at Cheriton took the engineers back to the drawing board for more headaches and calculations. Then normal tunnel-excavation will continue, the tunnel curving out to sea in a gentle arc and heading towards the sand dunes of France.

But, excavation will not be in one direction only, because different diameter machines will be needed for the subterranean and the submarine sections, and also because it would take too long. The phrase "time is money" might almost have been coined for a project of this scale, where a mere one week overrun could cost in the region of £5 million by 1993.

So three machines were to start their eight kilometre drive from Castle Hill and head towards the sea, but now, owing to pressure from environmental lobby groups, Eurotunnel has decided they will have to excavate the opposite way, from Shakespeare Cliff travelling inland.

Excavation will be led by the service tunnel machine, but it will not travel as far ahead as its equivalent on the submarine stretch. Spoil needed later for landscaping the terminal site will probably be taken along the completed service tunnel on a conveyor, from Shakespeare Cliff, where it will presumably have been stockpiled in the meantime. At the same time as tunnelling machines progress towards Sugarloaf Hill, the mechanical excavators will be tearing up a trench across the Coombe and the soil will be carefully stored so that it can be reinstated in the same order it was removed. The tunnelling machines could be transported across the Coombe on rails to start work again on another portal in Castle Hill, heading towards the first daylight Chunnel travellers from France will see in 1993. But it is more likely that this short but complicated stretch of tunnel will be excavated by a process called the new Austrian tunnelling method(NATM). NATM frequently uses a roadheader, a tunnelling machine used on many mine roadways, and then, instead of lining with concrete or cast iron rings, a concrete mix is sprayed all over the tunnel walls. Steel reinforcement grids could be attached to the tunnel walls, or steel fibres could be mixed in to increase the concrete mix strength. There will probably be several tunnels through at least part of Castle Hill, with an additional running tunnel cross-over under consideration. This portal, sited in an old landslip, will require careful engineering to stabilise it and ensure there is never a mud collapse on a train entering or leaving the tunnel. A borehole was sunk 70 metres through the edge of this hill, and many samples were taken for testing at Kingston Polytechnic by one of Britain's landslip experts. Special apparatus is used to carry out shear testing, to ascertain the rock and soil's history of movement and potential to move again. Ring shear tests were carried out at 0.25 metre centres down a 70 metre borehole to pick up residual strengths down through the Gault profile. The tests over six weeks cost about £15,000–£20,000. A reservoir uphill from the portal site will hopefully not add any extra surcharge to the soil.

Much of the coastline and some inland sites, too, are prone to landslides. A particularly dramatic one pushed a railway train off the tracks, parallel to the shore at Abbot's Cliff, Folkestone in 1915. Rubble flowed more than 400 metres out to sea and the damage to the railway took five years to rectify. Movements in the high cliff required re-routeing of the Folkestone to Dover road... if it happened again, it might save a little of the controversy over the A20 road improvement! Within a landslide's flow of the 1915 event, a heading to test a Chunnel machine in the 1920s lies partially concealed beneath another landslide, complete with its "inhabitant", the Whitaker machine. Eurotunnel may fund the removal of this machine. Railway tunnels along the Warren, Abbots Cliff and beneath Shakespeare Cliff are corkscrewing owing to movement in

Folkestone — a town preparing for change. Port-oriented business will adapt to rail freight and passenger demands instead.

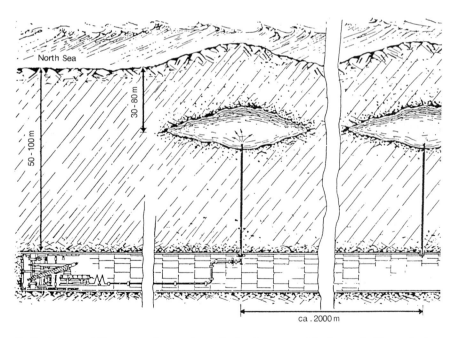

Chalk slurry could be pumped up to the seabed or into any voids close to it. The seabed topography would be altered but, with no property above, there would be no compensation bills to pay.

the chalk. There are frequent chalk falls, say every couple of years. In some cases the railway tunnels are seawards of the cliff crest. Sir William Halcrow and Partners is consultant to British Rail on this coast and also to Eurotunnel, so the information has been passed on! Dr Eddie Bromhead of Kingston Polytechnic said "Chalk is really weird. It is a high porosity material within large voids held open by calcite cement." Albert Hein , a Swiss geologist, has described Alpine rockfalls called Sturzstrom. A famous one near Elm involved dolomitic limestone flowing for several kilometres. To get this degree of mobility required a face 600 metres high... to get this in chalk said Dr Bromhead "needs a face 130 metres high or so... and it could run for three or four kilometres out to sea."

While machines are tunnelling towards Sugarloaf Hill, the others starting off in the old 1974 workings will probably be taken underground on the old inclined adit or new larger diameter one close to it. A new shaft will be excavated at Shakespeare Cliff as well. If the adits or shafts are not large enough, the running tunnel machine would have to be assembled underground in a large cavern called a machine chamber.

In 1974, the service tunnel machine was taken down in one piece, which accounts for the larger than otherwise necessary start to the heading. Since 1974 the UK Health & Safety Executive has increased in status and so in power. The dictates of this organisation, or requests of equipment manufacturers restricted in design by the confines of the Tunnel, have resulted in increased diameters for all the tunnels, not just since 1974 but also since the Channel Tunnel Group and France-Manche first drew up plans in 1985.

The UK Health & Safety Executive (HSE) requires a continuous walkway be provided during construction of the service tunnel. This meant an increase in the 4.5 metre diameter on TML's original proposal to the 4.8 metre diameter now agreed. The running tunnel diameter is solely dictated by the requirements of the rolling stock, not the HSE.

Some engineers involved in the project disapprove of the service tunnel walkway provision. They said "One was not needed in 1974. It was just accepted that men did not walk in the Tunnel and only travelled on personnel carriers". The engineers point out that, unlike many smaller tunnelling operations, everything will be fast moving underground, bringing materials in and taking spoil out, "even during construction it will be like a mini-metro, with trains travelling at speeds up to 48 kilometres an hour".

Ground pressures on the tunnel lining are greatest on the landward drive, and most particularly just as the tunnel passes beneath Shakespeare Cliff. Movements measured in British Rail tunnels show the whole cliff to be rotating. This torsional effect combined with the sheer mass of rock bearing down upon the tunnel will demand high strength tunnel lining.

What must ease the tunnellers' minds a little on this stretch is how well the 1974 heading has withstood such effects. Despite some flooding over the thirteen years, the circularity of the profile has been maintained — unlike the 1882 Beaumont Tunnel close by, which is decidedly elliptoid. TML has commented that it thought there was no flooding; a regular employee, presumably of the Department of Tranport, used the term, possibly not implying total flooding merely water ingress when pumping was infrequent.

Concrete and lining

The concrete lining segments of the 1974 tunnel are also in good condition, and were tested by TML in early 1987 to see what deterioration had occurred, if any. I had the good fortune to visit the site on 11 March 1987 with the House of Lords Select Committee. Men were test drilling through the front of the dismantled tunnelling machine. Remembering the concrete testing I had undertaken at university, I picked up one of the core samples that had been drilled out of one of several wall test stations and was reassured that, if nothing else, we *can* still make good concrete in the UK! A perfect blend of aggregate sizes, no voids or crazing... if I'd ever had doubts over the Tunnel's integrity, they were immediately banished for good. It was especially comforting in view of the "concrete cancer" that has dogged the construction industry in recent years, whereby silica in the cement or aggregate has reacted with alkali constituents and caused bridges to crumble away and dams in north Wales, Jersey and Scotland to leak.

Awareness of this problem led TML in 1987 to carry out extremely detailed aggregate testing in preparation for concrete segment manufacture at TML's new casting yard on the Isle of Grain in north Kent.

Colin Kirkland said "On the UK side the indications are that we have sound, hard Lower Chalk strata. It will be most expeditious to use an expanded lining and to grout if necessary".

"On the French side it will be extremely wet and friable. There will be segmental bolted lining within the tail of the shield, which would be completely watertight".

A range of lining types is available and all are capable of being erected by machines. Colin Kirkland of Eurotunnel said that this would be the high technology end of the project. The design of expanded concrete linings to their current status is a UK development. However, the machine to erect them is from Swiss designer–manufacturer, Robert

Scars on a landscape: the muddy path across a field to a drilling site on Castle Hill provides work for local contractors but a temporary eyesore as well.

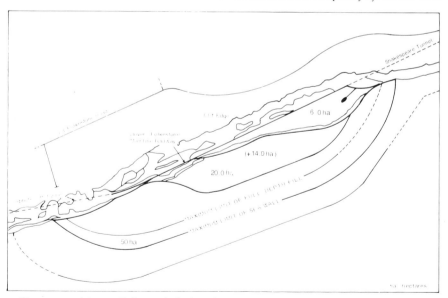

The extent of the spoil disposal platform has been one of the most controversial issues in the Tunnel project. In France, the chalk will be pumped in slurry form and and held behind a dam in an old dry valley until it one day resolidifies.

Waehlti. Any grouting required behind the lining will go through a butterfly-valve that will automatically close once there has been sufficient grout uptake. The lining is a development of that first used 25 years ago at Potters Bar in the UK. Segments rest one upon the other and the ring is expanded against the chalk by a key segment. Precast concrete wedgeblock lining will be used on the UK side, but bolted cast iron has been considered for the landward drive or elsewhere if problems are met.

Where ground pressures are higher, the tunnels will have thicker lining segments, but as the internal tunnel diameter must obviously remain the same, this will require a larger void to be excavated and larger diameter tunnelling machines. Bolted concrete segments have been specified for the French side, for at least the first five kilometres of tunnel. Formwork has to be made with notches for some segments and several rings-worth of segments will be available in the tunnel. In the unlikely, but not impossible, situation of concrete segments being broken, delay will be inevitable, as they are brought in in the correct order on segment bogies.

Various companies hoped to bid for segment manufacture before TML decided to set up its own casting yard on the Isle of Grain. Charcon, now owned by Costain, in turn one of the member groups of TML, tried to bid separately and Costain tried to bid in a new consortium with CV Buchan. In early 1986, Charcon was producing 1000 tonnes a week of concrete. If it had gained the service tunnel contract, it would have had to upgrade to 1000 tonnes a day. Mr Rupert McBean, managing director of Charcon said "There is no suitable aggregate in Kent, and it would take 40 to 50 acres of land for a factory to build these segments and up to 1000 extra staff".

Cast iron tunnel lining segments

TML initially issued a tender document for the supply of about 80,000 tonnes of cast iron tunnel lining segments but met a poor response. Some said this was because the document phrasing was too loose and there was uncertainty all round about programming. TML's comment was that no manufacturer in Europe had the capacity to supply all the linings to the programme at an economic cost. So, it said, "the contract was then broken down into smaller contracts so that competitive bids could be obtained". The present pricing policy of cast iron lining manufacturers indicates that they are two to three times the cost of concrete linings. At present cast iron linings will only be used for special applications where it is impracticable to use concrete linings, in total about 75,000 tonnes for the UK and 5000 for France. Those remaining in the 1974 heading at Shakespeare Cliff appear to be in a far worse state than any concrete present. However, the concrete had been strengthened because British

Rail feared the abandoned tunnel could affect the stability of their rail tunnels above.

Unplugging the Channel

If a fault zone is connected directly to the sea-bed, it provides a passage for water inflow. If the tunnelling machine dissects a fault, the Channel could be unplugged. The water would not just seep in, but would be under high head, the pressure being proportional to the depth of the sea at that location. The same phenomenon could occur if an old borehole is encountered; water under 12 bar pressure would flow into the tunnel, with sea level as much as 40 metres above the sea-bed. High pressure could enlarge the hole or plane along which the water flows and gush into the tunnel at such a rate that the one minute required to push out the steel blade annulus-filler might not be enough. Probably the worst scenario is a high inrush of water and sediment, flowing from near the tunnel machine picks, along the outside of the machine, as far as the machine driver and other men employed to ensure safe segment erection. The annular blade might not be pushed out quickly enough, or it might be jammed in its housing, not having been used regularly, if at all. Even if all this were to happen, all would not be lost. There are not many men working close to the machine head, and hopefully — unlike the ill-fated pumping station at Abbeystead, England, where 16 people died in 1985 — there will never be large groups of visitors to this "danger zone". With such an inrush, three men might drown, and the laser, tunnelling equipment, and segment erection facilities be damaged. But with all the precautionary measures being taken, and the experience of the engineers involved, it is about as likely as Margaret Thatcher joining the Communist party! Such accidents have occurred elsewhere though, but not involving loss of life, merely great cost.

For a 3.6 metre diameter power station cooling water tunnel from the Isle of Grain to the Isle of Chetney, Kent, the tunnellers struck an unfilled borehole, and so damaging was the inrush that the shield was abandoned and the tunnel diverted around it and the ground had to be frozen with liquid nitrogen to continue working. This will not be possible with the Channel Tunnel. Problems encountered will have to be solved, not avoided, because the alignment in all three planes is too vital. Realistically, whether a fault or a borehole, when water rushes in, it will suck sediment with it that will fill the offending orifice in a matter of minutes. Then tunnellers will retract the machine a little and carry out comprehensive chemical or cement grouting to fill any voids present.

Before the current project, a major drilling exercise was last carried out
in the mid-1960s. Excessive punctiliousness by civil servants, said Sir
Harold Harding, delayed the work programme until the winter months.
"Ships had to return to harbour leaving 250 millimetre lining tubes bored
as guides into the sea bed. On returning several tubes were found to have
broken and collapsed." The mechanism of failure and recovery of
abandoned holes was a complicated process. It is not certain that all
boreholes were located and sealed. Probes of 100 millimetre diameter are
not likely to find old boreholes but every probe will be injectable and *tube à
manchette* grouting will take place from the rear of the machine. As
probing cannot be continuous, there is still a chance of striking an old
unfilled borehole.

Chalk may be taken by conveyor along a service tunnel to the terminal site at Cheriton.
Muck cars are easier to maintain but conveyors can transport the material up shafts as well
as inclines.

Grouting

Over 80 per cent of the Tunnel will be in Chalk Marl. In France they expect
to grout difficult ground before boring the tunnels. Probing 100 metres
ahead of the face after every seven working shifts will enable the team to
detect and treat any fissures found. Where ground is really waterlogged,
it may be tackled in several ways. It is possible to drill long holes, circulate

freezing brines through them and so freeze all the water in the surrounding ground. This is very expensive and is unlikely to be viable or necessary beneath the Channel, but grouting will be carried out as a matter of course. And, there will be narrow probes extending one hundred metres or more in a peacock tail fan from the front of the service tunnel, the probing exercise taking place every day for an hour or so while maintenance is undertaken on the boring-machine. Should waterflow from one of these probes exceed one litre/minute, grout will be mixed and injected into the surrounding rock and then tunnelling would continue, to within twenty metres of the end of the last probe, thus always leaving a good safety margin of "known" geology ahead of the machine.

Probing from the machine

When the "running " tunnels are excavated , there will also be probes ahead and around them. Hopefully these will totally eliminate geological surprises, but, should there be an unfilled borehole undetected in geophysical surveys, the chances of a narrow probe bisecting a narrow borehole are also slim. The position of all boreholes should be plotted on charts when they are drilled, and each hole should be capped or filled after all sampling and testing are carried out — but the integrity of the drilling contractor or crew so far in the past cannot be guaranteed.

Drainage

There will be four pumping stations within the Tunnel complex, one very close to each shore and another one about four kilometres from each shore. In an extended W form, the service tunnel will have gradients of 1:100, 1:1000, 1:1000, and 1:100. The pumping station within the 1974 workings has six pumps. Until construction work restarts in earnest, only one is used about once a week.

Tunel-Français

Tunnelling from the French side will not be able to make use of their 1974 workings - or indeed those of any earlier attempts. All they will have gained from these escapades is knowledge of problems they are bound to encounter.

Excavation of a 55 metre diameter, 120 metre deep shaft started in France at the end of February 1987 following completion of a one metre thick diaphragm wall which goes 20 metres into the ground. The diaphragm wall, as part of "temporary" works, is of a far greater diameter than the central shaft.

Construction of the enormous shaft will form the working site from which all six tunnels will start — three going landwards a mere 3.2 kilometres and three heading off under La Manche. The Hybrid Bill would not have permitted such a structure to be started in the UK until Royal Assent had been received. Fortunately for the French, and indeed for the whole project, their definition of "temporary works" can be applied rather loosely when necessary. Without the facility of old, usable workings and with much more difficult geological conditions for at least the first five kilometres, it was vital that the French have a head start.

Much fuss was made in both Houses of the British Parliament about the relative quantities of chalk spoil to be deposited in each country, with no-one really seeming to explain in sufficient detail why in fact the French were getting the worse end of the deal. As Eurotunnel is a totally bi-national company it does not matter that tunnelling machines for the French tunnels cost about three times as much as those needed for the UK side, but the French will have more dangerous working conditions and, although less spoil, it will be in a particularly difficult form to handle. French tunnels have to pass through Upper Chalk, which is quite fractured and fissured, each little joint being able to transfer water rapidly through the rock. So, although they will use full face tunnel boring machines of similar dimensions to the UK side, the TBMs will operate in a very different manner and will have a totally sealed cutting system with complete exclusion of water. However, when conditions improve further under the Channel the machine will be adapted accordingly.

Instead of picks, the machines will excavate with roller cutters; these are discus-shaped, of about 305 millimetres diameter, made of metal and impregnated with diamonds or other cutting media all round the periphery.

Although the Robbins machine built for French use in 1974 has been well-maintained, technological advance in the intervening years has determined against its use. SITUMER had ordered an articulated 4.95 metre diameter tunnelling shield from Robbins of Seattle in the USA, which was assembled in Sangatte but, owing to cancellation of the project, was never lowered down the shaft and tested in the conditions for which it had been designed.

Model number 165-162 weighed 245 tonnes, had a torque of 109,000 kg.m, thrust of 471,000 kg and a cutter head powered at 615 kW. Its 30 discs and five central drag bits are still fresh from the factory and are never likely to cut any rock.

Like drum brakes on cars, grippers were made to push out from the shield against the tunnel wall providing thrust for jacks to propel the machine forward. This system does not work well if the rock caves in.

The roller cutters were arranged on six radiating ribs with scoop-like arms interleaved to shovel the muck away. It would have been possible to swop the roller cutters for picks if ground conditions had warranted it. Segments could be placed at a slower rate (or faster) than the main tunnelling speed, as the erector travelled on a long beam behind the machine.

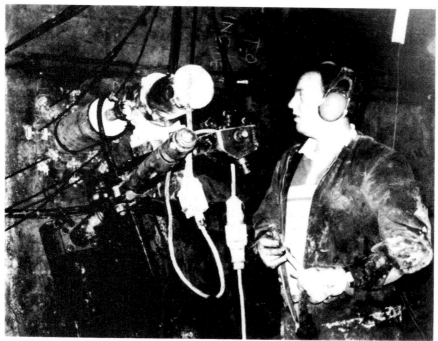

Testing concrete quality in 1987 in the abandoned service tunnel beneath Shakespeare Cliff. The concrete has withstood the effects of time with no visible deterioration. Source: Mott Hay and Anderson.

The American company Robbins again won the first French TBM order, in December 1986. Their new design — in collaboration with Japanese company Komatsu and French manufacturer Som-Delattre of Valencienne - still operates on the roller cutter principle but will now have a closed face. This means that the machine will employ a slurry circulation system, whereby the excavated chalk is mixed at the face with fresh water, and the rock is then pumped away. Fresh water is used at this location to decrease the salinity of the spoil, which is to be dumped on top of the aquifer that supplies drinking water to nearby settlements. Once contaminated, it is very difficult, if not impossible, to reinstate an aquifer.

Bentonite may be used to provide temporary ground support; if the machine stops cutting the clay cakes on the surface, when the machine starts again the waste cuttings and clay flow out in a homogenous mixture.

A circulation plant on the ground surface separates spoil from bentonite and then pumps the clay back to the face. This system eliminates the need for workers to enter a compressed air environment and, as with the Howden machine, the cutting tools can be changed in the comparative safety at the back of the cutter head.

In order to tender for TBM design and manufacture on the UK side of the Tunnel, the Robbins company formed a joint venture with Markham & Co. of Chesterfield. They issued a statement in February 1986 saying that they intended "to pursue the Channel Tunnel Project and will be presenting a united tender for the supply of boring machines and back-up equipment". The venture was not successful for the first two machine orders placed in the UK.

Robbins have their own segment erection system as well as their own design of muck wagons. The early part of the French tunnels at least is likely to be lined in cast iron. This is partly due to tradition and partly due to the material's superior ability (when compared to concrete) to deal with the soil conditions expected in France.

Five tunnelling machines will be used for the French side, although these will cost considerably more than the six UK TBMs. One machine will drive the service tunnel landwards and one will drive it seawards; two machines will drive both running tunnels out to sea and one machine will drive a landward running tunnel away from the shaft and at its portal will be turned around to excavate the other short length of running tunnel.

Mr Ian Elliott, an engineer with worldwide tunnelling experience, is employed by Mott Hay and Anderson, consultants to TML. He said "The Robbins machine is all-singing and all-dancing... its supposed to be able to do almost anything." He believes that equipment for the Tunnel is similar to much that has been used on the Cairo Wastewater project, where British consultants and contractors have played such a major role.

Another UK engineer said he thought that "our tunnel will be about half the cost of theirs" - implying not just that the French had more difficult ground conditions but that they were spending in excess of their needs.

Soletanche are building the parrarie moule ... the diaphragm wall of the access shaft; Guitoli and Colas have the platform puits contract (the shaft platform and site preparation); Transloko has supplied site buildings and Wattez had a small contract for electrical installations.

Spoil disposal on the French site involves construction of a dam across a dry valley, and pumping chalk, as a slurry, behind it. On aerial photographs much of the ground surface appears pock-marked... some hollows may be natural craters but some are undoubtedly the result of pounding by bombs in the last war. But, perhaps not all the bombs exploded. "Wouldn't excavation of the dam foundations be rather hazardous..?" asked one British engineer. "Oh..." the French engineer paused.. then beamed "Well, the Japanese will expect a contract or two..."

Robbins machine

The machine will be manufactured in joint venture with French company Som-Delattre and West German company PWH. Som-Delattre have previously manufactured tunnelling shields for other European projects. This multi-purpose machine is the first of its kind in the world of tunnelling. Robbins have decided to retain the traditional roller cutters but to use the most up-to-date Japanese technology for ground support. The contract, worth in the region of $16.5 million (£11 million), is for a machine that will bore 16.27 kilometres. It was completed in October 1987.

Robbins, Becorit and Muhlhauser also tendered for the construction rolling stock. One competitor in this contract race said "they're not high technology, just ..... boxes on wheels".

The Robbins TBM supplied in 1987 was fitted with a number of special features. It will work in open mode when there is no water pressure at the heading and in a closed mode when water pressure develops. In the open mode the cutterhead advances while the shield remains stationary, allowing the tunnel lining to be installed as the heading is bored. In the closed mode, the cutterhead and shield advance together, thrusting off the tunnel lining. In this mode the lining is installed between advance increments. Special seals are used to maintain water tightness under high pressure. The machine will come equipped with three muck discharge systems and a screw conveyor. When operating in the open mode, muck discharges through a free gate in the cutterhead via the screw conveyor to a flight conveyor and finally a belt conveyor, which carries it aft to muck cars.

If the closed mode is required in the first two kilometres of boring, the free gate will close and a slurry discharge system will be used to pump the muck through a slurry line to the shaft. Beyond two kilometres, if the closed mode is required, piston dischargers will pump the muck through pipes to the muck cars. Robbins has collaborated with Komatsu of Tokyo, Japan, on the pressure shield aspects where Komatsu has special expertise and experience. The Robbins machine is expected to advance at the rate of four metres an hour in the chalk, achieving penetration rates of over seven (7.2) metres an hour. In total the machine will bore 16.27 kilometres of the undersea service tunnel and is scheduled to finish by 1 September 1990. The invitation to tender for construction of tunnelling machines to be used from the French side appeared in the Official Journal of the European Communities No S 90/55, on 13 May 1986. Transmanche called "for design, manufacture, shop assembly, testing, delivery to site, assembly, commissioning, maintenance and operational support, including efficiency guarantee, of tunnelling machines for working and lining the French part of the Channel tunnels".

It was a tall order, going on to specify that the machines should be able to function in almost any mode or be able to deal with any eventuality! Spoil removal in confined operation, hydraulically, or in open operation , by train; lining speed over a six hour period with a 15 minute lining ring cycle, of four metres an hour in the small tunnels and almost five (4.7) metres an hour in the larger tunnels; integrated equipment for exploration and treatment of strata during tunnelling; hydrostatic pressure at standstill of eight or 12 bars, in motion of five or seven bars.... and in addition to all the technical details the companies had to prove their financial stability in recent years and show capacity to manufacture at least two machines at once, within ten months.

Komatsu listed the required capabilities as follows:

Maximum boring speed 12 centimetres a minute; concurrent segment installation and boring; segment assembly within 15 minutes a ring, boring of uniform chalk layer (muddy limestone) and degenerated chalk layer containing hard flint pebbles; open shield boring in good ground and closed shield boring in faulted and fractured zones, and shift capability between these methods in a short time; withstand maximum water pressure of 12 kilogrammes a centimetre squared and work under maximum water pressure of 9 kilogrammes a centimetre squared; probe boring from within the machine.

The main specification for the marine service tunnel machine:

Boring diameter 5.6 metres; shield jacks gross thrust 4000 tonnes; shield jack speed 15 centimetres a minute; cutterhead thrust 1200 tonnes; cutter head stroke 1.45 metres; machine length 10 metres; electric motors total output 1400 kilowatts; weight 450 tonnes.

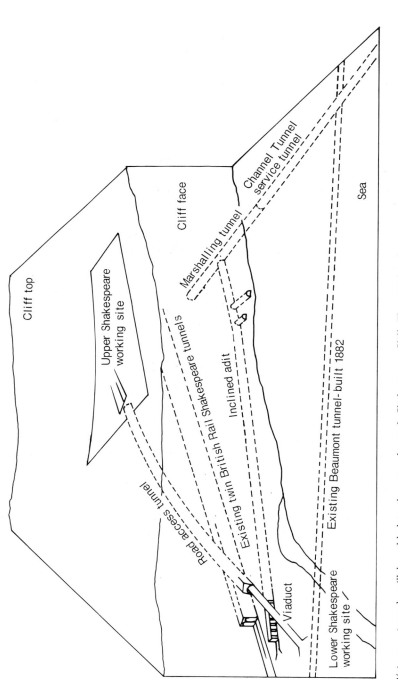

Yet more tunnels will be added to the maze beneath Shakespeare Cliff. The weight of the cliffs and the complicated loading pressures traversing from high load to low load as the tunnels progress seawards, make computers essential to calculate the forces that will act on the tunnel linings.

Cliff top

Cliff face

Upper Shakespeare working site

Channel Tunnel service tunnel

Marshalling tunnel

Inclined adit

Existing twin British Rail Shakespeare tunnels

Existing Beaumont tunnel - built 1882

Sea

Road access tunnel

Viaduct

Lower Shakespeare working site

Safety and ventilation

Hinged doors on cross-cuts; louvres in doors set to balance air flow, with larger openings towards the centre of the tunnel. Tunnel ventilation is complex.

Forced ventilation from each shore should theoretically balance in the middle, but contractors are likely to install some kind of baffle there. The doors will be electrically locked so that in effect they only function one way and personnel in the service tunnel could not walk into the path of a fast train. Normal ventilation is required to provide fresh air for the occupants of the running and service tunnels. Emergency ventilation is to cope with short-term excesses of temperature or pollution in the main tunnels. Aerodynamic drag constitutes about 90 per cent of the total train resistance. Thus it determines to a major extent not only the power requirement of the locomotives but also the heat input to the tunnel. Drag can be modified by pressure relief ducts. Air pressure fluctuations generated by the train movements in the tunnel need to be small enough to avoid discomfort to passengers and railway personnel with acceptable airflow velocities maintained in those parts of the Tunnel complex used by personnel. The ventilation flows need to provide an acceptable environment in terms of temperature, humidity and pollution levels for the Tunnel occupants under the vast majority of normal and abnormal operating conditions. In the event of a fire it is necessary to prevent the entry of smoke into the service tunnel and to restrict the movement of smoke towards people during disembarkation of the train and evacuation into the service tunnel.

The amount of specialised ventilation and power supply equipment needed, shows that although the TBM is the heart of the operation, everything else is vital to its success. Some mining restrictions are applied in the tunnelling environment, but it will not be essential for everything to be flameproofed or "intrinsically safe", terms which related to equipment's ability to not catch fire if there is an electrical spark, or not to spread flames to other equipment. Methane is not anticipated (although peat has been found in Channel boreholes), but monitors will check air quality continually — from the point of view of explosive levels and also to check the air is being circulated efficiently such that workers have a relatively pleasant and healthy environment. The cooling of tranformers and generators raised some controversy in the UK when it was discovered that equipment used in the 1974 attempt was cooled by polychlorinated biphenyls (PCBs) instead of diesel oil. This in itself is not unusual — use of diesel equipment in tunnels is avoided because of safety from fumes and fire in such a confined environment. But, when contractors were called in to dismantle most of the Priestly tunnelling machine and other equipment remaining in the tunnel, so that the metal could be sold for scrap, they were not informed — or did not follow instruction — that highly toxic chemicals were contained in the transformers.

PCBs are among the most noxious substances known to man, causing kidney and liver failure if they enter the respiratory system, and birth defects are suspected but not yet proven. PCB disposal will probably be handled by one of only two UK companies that are suitably equipped, Rechem International based in south Wales. Mildly contaminated soil will be sent to a licensed tip a few miles away, at Richborough. The spillage seemed to be unnoticed until the site was checked by Eurotunnel/TML before starting preparatory work. Fortunately for the consortium, the £1 million disposal bill will have to be met by the UK Government. Scientists from the Southern Water Authority visited the 1974 heading in March 1987 and collected water samples from the pool beneath cast iron tunnel lining, just behind the remnants of the Priestly machine. They also took samples from the pumping station closer to the shore, to see how far PCB contamination had progressed, if at all.

A process developed by Westinghouse, in which PCBs are destroyed with high temperature plasmas, may ease the problem. If the material is burnt in ordinary incinerators and combustion temperatures are too low, highly toxic dibenzofurans and dioxins are formed instead. But if air is passed through an electric arc at temperatures above 5000 degrees Celsius, molecules in the air are disassociated into electrons and positive ions. This plasma then breaks down molecules passing thorugh it into atoms and ions. These recombine on cooling to form simpler compounds. Thus PCBs can be converted into carbon monoxide, hydrogen chloride, nitrogen and oxygen "waste products". The carbon monoxide could be used as a fuel or flared-off as gases on oil platforms are; absorption in sodium hydroxide solution neutralises the hydrogen chloride. Efficiency of 99.9999999% is achieved, and perhaps most usefully, all equipment can be stored on a 14 metre long trailer and the disposal carried out on the contaminated site.

Eurotunnel had to buy the access tunnel and the 1974 heading from the Government, complete with a stockpile of concrete lining segments just dumped on the site thirteen years ago. The purchase was delayed while the solution to PCB disposal was determined. The ugly grey segments are an eyesore that local environmentalists seem to have been oblivious to while no Channel Tunnel project was on the drawing board, and the broken concrete wartime reinforcements all along the cliff top also seem accepted as part of the English heritage with no judgement passed on their aesthetic value — or lack of it.

Construction materials

Movement of bulk materials to the various construction sites caused considerable concern to environmental and other protest groups. Materials needed include minestone (colliery shale), quarry waste and marine-dredged fill as well as aggregates, sand, road or terminal sub-base, railway ballast, drainage materials, pre-cast concrete tunnel lining segments, cast iron segments, reinforcing steel, asphalt, cement, piles for the sea wall and large rocks to protect it from wave attack. Once railheads are established, most of this material can be moved by rail, but initially there will be a large increase in lorry traffic on local roads in the Folkestone and Frethun areas and, depending on the source location, much material will be brought in on coastal shipping routes. High strength aggregates may be obtained from crushed rock sources in Kent or imported from elsewhere. There will be a lot of pressure on southern England's supplies with several other major construction projects underway at the same time, including a new crossing of the River Thames, a new airport "Stolport" and other projects in the London Docklands as well as much secondary development associated with the Tunnel. Cement may be imported from Greece or supplied locally; sands and filter layers will come from Dungeness and Conningbrook; reinforcing steel from south Wales and Sheerness; asphalt will be manufactured at Hothfield; concrete segments will come from the Isle of Grain; cast iron segments could be sourced in the English Midlands, France or West Germany, but the bulk of the materials will — as shown in this list — come from areas within Kent. The French will also minimise transportation effects on communities by looking for as many localised suppliers as possible.

Man or mole?

Drivers talk of tunnelling machines as car drivers would describe their vehicles, or captains their ships. They have an affinity for the rock and for the powerful motive force of the machine that cuts into it. It is a mole-like existence going ever deeper into more perilous territory. As important as the machine's progress through rock is its lining capacity and spoil-removal rate. The rock is cut, removed, lining erected and the machine moves on. But although cyclic, the demand is for the machine to cut continuously and the other factors to keep pace. This is exceptionally mechanically demanding of any machinery but particularly so when the machine is also expected to work a seven-day week, 365-day year.

Far more than any road, bridge, or tower block construction, tunnelling has that exciting element of danger... Few people outside the industry ever get to see what happens underground.. the dirt and dust, the grinding of the machine, and when the machine stops, the wagons are gone, the workers are still... then the unnerving sound of the rock itself, settling onto the lining rings and creaking in grudging acceptance of the intrusive barb through its very core.

Lower Chalk, with its high clay content, reacts slowly and quietly to removal of a tunnel's worth of rock. The only noise likely to concern Channel tunnellers is that of water dripping, trickling or gushing through.... hopefully there will not be enough even to make the concrete damp.

Organisation has to be meticulous. There have to be sufficient stocks of lining segments on site to allow for strikes, bad weather, train derailment or even a dramatic increase in tunnelling speed, but not so many segments that the site is too cluttered to move other equipment on it. A steady production line, frequent deliveries and constant rate tunnelling is all that is required.

TENDER DETAILS

The original tender documents sought bids for all six UK tunnelling machines. Lack of finance or other deliberations meant that TML missed the three month validity date of bids on more than one occasion and had to ask manufacturers to resubmit the document required:

Contractors keep on tunnelling until the two sides meet, irrespective of where they might be. In that way, any delays due to unexpectedly adverse ground conditions, strikes or other causes are minimised. Once the tunnelling machines are well on their way under the Channel, apart from a few close encounters with the Gault Clay beneath the Lower Chalk and whatever unpredicted geological hazards are yet to surprise them, the construction story will be quite unexciting until the French and British tunnellers can hear their counterparts' machine rumbling through the rock towards them. There are a few areas of faulting and folding fairly close to the shores, but it is not expected that these will cause problems. Machines will of course, break down... though not necessarily all of them if the carefully planned maintenance schedule is adhered to. But on the best-planned project, mechanical failure cannot be avoided - merely its effects minimised. So, if a main bearing shaft shears in two, the similarity of spare between machines should allow vital parts to be kept in stock.

Starting the Tunnel in January 1974. 1987 should see further progress.

Rocker picks chip out the chalk.

It is virtually impossible to draw comparisons of tunnelling rates with other projects because of the many variables including ground conditions, tunnel diameter, labour skills and shift working. Tenders were invited for TBMs with a maximum design rate of advance of 1800 metres a month. In the contract programme it is assumed that the actual rate achieved will only average 1000 metres a month in good ground and considerably less near the French coast. Generous leeway has been built into the programme to allow for delays.

On the landward drives the natural rock outcrops in combination with site investigation, will have given a very detailed picture of the conditions to be met. In fact such a detailed picture has been achieved that environmentalists asked for the cut-and-cover tunnel to be moved, to avoid a "site of special scientific interest" - namely a small and localised geological deposit. There are penalties of accuracy!

Provision shall be made to accommodate trains of muck cars up to 140 metres long; the manufacturer shall co-operate with the rolling stock supplier to ensure compatibility of systems; the minimum installed capability of each pumping system must be 100 litres a second; duplicated systems of pumps to maintain a water and slurry free face shall be fitted as a permanent unit with suction pickups close to the cutterhead.

A. One marine service tunnel machine; 5.10 metre diameter, 25 kilometre drive.

B. Two marine running tunnel machines; 8.06 metre diameter, 25 kilometre drive

C. One land service tunnel machine; 5.46 metre diameter, 10 kilometre drive,

D. Two land running tunnel machines; 8.46 metre diameter, 10 kilometre drive.

Dimensions may vary slightly as the lining design is fully developed.

Delivery dates shall be as follows ... and form the essence of the contract.

A: May 1, 1987; B1: May 1, 1988; B2: September 1, 1988; C: July 1, 1988; D1: October 1, 1988; D2: February 1, 1989.

Erectors must complete one full cycle in the running tunnels in 18 minutes and the service tunnel in 13 minutes.

Suitable jacking systems are needed to force the lining locking key pieces into position.

Segment erectors shall be equipped with an axial float and be capable of erecting a ring immediately behind the machine body or the preceding ring.

Ring length of lining shall be 1.5 metres.

Erection, boring and mucking-out shall be simultaneous.

Machines must be able to handle both spheroidal graphite cast iron or precast reinforced concrete segments.

When Colin Kirkland and Jean-Loup Dherse were asked what memorial they would like in the Channel Tunnel for themselves — and reminded that Isambard Kingdom Brunel, the great Franco-British Victorian tunneller, built a tunnel through which the sun shone only on his birthday each year—Monsieur Dherse said they would have to install mirrors to achieve such an accolade for both of them. Kirkland, in his disarmingly modest way, said that the Tunnel would be monument enough for *all* those whose hard work made the scheme possible.

References

Geological results of the Channel Tunnel site investigation 1964–65, Report 71/11 British Geological Survey, JP Destombes and ER Shepherd-Thorn.
Handbook of Mining and Tunnelling Machinery, Barbara Stack, Wiley.
Henry Marc Brunel: The First Submarine Geological Survey and the Invention of the Gravity Corer, Desmond Donovan, Elsevier.
The Stability of Slopes, EN Bromhead, Surrey University Press.
Tunnels: Planning, Design, Construction, T.M. Megaw & J.V. Bartlett. Ellis Horwood.
Tunnelling History and my own Involvement, Sir Harold Harding, Golder Associates.
Geology of the Country around Canterbury and Folkestone, J.G.O. Smart, G. Bisson, and B.C. Worssam, HMSO.
Various issues of *World Construction, Tunnels & Tunnelling, New Civil Engineer*, and other periodicals.
British Standard Code of Practice for Safety in Tunnelling in the Construction Industry, BS6164: 1982. One of the cooperating organisations was the Department of Transport.
Aerodynamics & ventilation of a proposed Channel Tunnel, D.A. Henson of Mott Hay & Anderson and R.G. Gawthorpe of British Rail. Paper presented at a *Symposium on the Aerodynamics & Ventilation of Vehicle Tunnels*, 23-25 March 1982. [It refers to a single running tunnel option.]
Hallsands, a Pictorial History, K. Tanner and P. Walsh, Sugden Design Associates.

4

Railway of the Century

Murray Hughes

There is a school of thought that believes railways are outmoded, inefficient and doomed to decay. Certain parts of British Rail would appear to support this hypothesis, and at times I find myself tending towards the same opinion. Yet it is not really so. A survey in the January 1987 issue of *Railway Gazette International* revealed that new railways totalling more than 70,000 kilometres are planned or under construction, with actual work in progress on 14,500 kilometres in more than 50 countries.

The grand total included 25 kilometres of new railway planned in the UK and a further 25 kilometres in France. This represented the Channel Tunnel, which by autumn 1987 will enter the main construction phase.

Make no mistake — the 49.4 kilometre Channel Tunnel is a railway. Not an ordinary railway, but a sophisticated link welding together the national rail networks of Britain and continental Europe. It will form a strategic connection in the pattern of European inter-city services and open up opportunities for railways to recapture long-distance freight traffic lost to the lorry.

A railway under the Channel means that a secretary in Maidstone could decide on the spur of the moment to spend her free day shopping for summer clothes in Paris. Catching a morning local train to Ashford and changing to an international express, she would be in the French capital by lunchtime. Admiring the latest creations by Pierre Cardin at Au Printemps and Galeries Lafayette in the afternoon, she could catch an early evening train back to Ashford, treating herself to dinner en route, and arriving home just three hours after leaving the Gare du Nord.

In terms of journey time, Paris will be no further from London than Newcastle is at the moment, so a stress-free day trip on the equivalent of the *Tees–Tyne Pullman* will spare the businessman negotiating a deal with a French contractor the frustrations of getting to and from Heathrow and Roissy-Charles de Gaulle airports. The school party from Manchester heading for a week's skiing in the Alps will no longer suffer the herding, hassle and sordid squalor of crowded Channel ferries in bad winter weather thanks to a special holiday train to Lucerne.

The wine merchant from Cologne who likes fly fishing might take his family on holiday to Scotland. A direct train to Edinburgh would eat up the 1 215 kilometres in not much more than 8 hours, which is shorter than today's fastest train service from London to Cologne via the Dover – Ostend Jetfoil. All this is just six years away.

As well as a primary railway artery, the Tunnel will simultaneously form a 'rolling motorway' joining the road networks of Britain and France. Here lies the key to the Tunnel's profitability — it will carry rail and road traffic on a single infrastructure. Once built the Tunnel will have minimal operating costs and it will quickly start to turn in a profit. Eurotunnel predicted an 11 per cent real rate of return on its shares in its prospectus for Equity 2 (Chapter 1).

Contrary to popular belief, not all railways run in the red. In Japan a number of privately owned commuter railways operate in the black. Thanks to an imaginative fares' policy with tickets such as Travelcards making it easy for people to use public transport, even the London Underground now covers its operating costs from fares' revenue.

The giant railways in the USA are run for profit too. The 21,000-kilometre Conrail network, for instance, formed in 1976 from a clutch of bankrupt railways as a government-owned organisation, was sold on Wall Street for a staggering $1.6 billion in March 1987 after consistently turning in a profit since 1981.

Most famous of all money spinning railways is the Tokyo – Osaka – Fukuoka bullet train route in Japan that whisks 128 million passengers a year through the teeming conurbations on the coastal belt of western Honshu at up to 220 kilometres per hour. This remarkable inter-city railway became a linear gold mine almost immediately after it opened in 1964, and so it remains today.

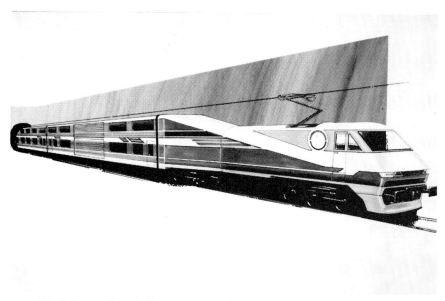

Artist's impression of a Eurotunnel shuttle train; the aerodynamic styling is to reduce drag effects due to the length of the Tunnel and the air pressure that will build up in front of the train.

Nearer home, France's high-speed line between Paris and Lyons that opened in two stages in 1981 and 1983 is another profitable railway. Sporting the world's fastest trains which cruise the 427 kilometres between the two cities in just two hours at up to 270 kilometres per hour — it has been so successful that other cities in France are clamouring for their own high-speed services. The Fr8.5 billion cost of this line and the Fr6.4 billion needed to buy its fleet of bright orange *Trains à Grande Vitesse* (TGVs) had to be funded by French National Railways (SNCF) on the open market.

In 1985 SNCF earned FFr3.5 billion from TGV services, of which FFr1.5 billion was spent on servicing the debt on the original investment. Working expenses amounted to FFr1.2 billion, leaving a clear profit of FFr800 million for the year. SNCF confidently predicts that all capital and interest charges will be paid off by 1990, less than ten years after the opening.

In the 1990s, Eurotunnel will join the elite among the world's railways that operate in the black, becoming the vital link on what promises to be Europe's busiest inter-city route. The Tunnel will consist of two single track rail tunnels, each with an internal diameter of 7.6 metres and a central service tunnel 4.8 metres across. There will be cross-passages and pressure relief ducts at regular intervals (Chapter 3). Crossovers sited near the Tunnel portals and at two intermediate locations under the sea will permit sections of the Tunnel to be closed for maintenance (or if an accident occurred) without interrupting the train service.

The diameter of the running tunnels was dictated by the need to carry lorries and coaches up to 4.2 metres high on the rolling motorway trains, while still allowing sufficient clearance for the 25 kV 50 Hz catenary from which trains will draw electric power.

The concept of a rolling motorway under the Channel dates back to the last Tunnel project that was unilaterally aborted by Harold Wilson's Labour Government on 20 January 1975. It was exactly 11 years later that President François Mitterrand of France and Britain's Prime Minister Margaret Thatcher signed the historic Lille accord launching the present Channel Tunnel scheme. Apart from the fundamental difference of private instead of public finance — a product of the British Conservative Government's rigid ideology — the current scheme is practically a carbon-copy of the last.

Studies in the 1970s proved conclusively that combining the roles of rolling motorway and through rail link in a single infrastructure produced the best financial return. When the project was examined again in 1983–84 by a group of two British and three French banks, they concluded that "the only proposal which is both technically acceptable and financially viable from a banking viewpoint is a twin-bore rail tunnel".

Not that this inhibited the usual rash of bizarre ideas when the French and British governments invited promoters in April 1985 to submit proposals by 31 October in the same year for a "Channel fixed link". These included a mind-boggling scheme by a Mr Boothroyd of the Boothroyd Airship Company who envisaged building a cross-Channel railway bridge with the help of heavy-lift airships. When this and the other proposals (one of which predictably included fairy tale castles in mid-Channel) had been winnowed out of the competition, the Channel Tunnel–France-Manche group that later became Eurotunnel was left as the winner.

None of the other schemes could hope to achieve the same rate of return, as they all needed a separate infrastructure to perform the dual roles of motorway and rail link. The hastily cobbled Channel Expressway plan by James Sherwood of Sealink British Ferries to build a road tunnel with a rail track down the middle was not adequately designed or costed, although it was taken quite seriously by senior members of the British Government, who were obsessed with the idea of a drive-though link.

Running through trains between Britain and the continent is nothing new. The gauge of both railways is the same, although the loading gauge (the profile to which trains must conform to pass under bridges and past lineside obstructions such as platforms) is smaller in Britain than the so-called Berne loading gauge which is standard on the continent.

Through rail services

Train ferries first plied the Channel between Southampton and Dieppe in 1917, and after the First World War a service began between Harwich and Zeebrugge, which continued until 1987, when British Rail concentrated all its train ferry traffic on Dover. About one million tonnes of freight a year travels in a fleet of around 7000 wagons specially built to fit the British loading gauge. They carry all kinds of goods ranging from chemicals to steel coils, grain, wine and china clay.

Glamour came to the train ferries in 1936 with the inauguration of the legendary *Night Ferry* sleeping car service between Paris and London. It was interrupted during the war, but resumed in 1947. Ten years later a Brussels – London portion was added to the train, which throughout its life exerted a special fascination for those who were privileged enough to use it, among whom were many carriers of diplomatic bags.

Air travel began to sap the *Night Ferry's* clientele in the 1960s, and a long, slow and painful rundown took place, culminating in the train's withdrawal in October 1980. It was a sad day, ending the possibility of a comfortable overnight journey from London, which permitted a Paris arrival before 09.00.

Since then, business travel has deserted the train–ship–train mode, partly because of the length of time the journey takes, but also because of the blatant disregard that the ferry operators have for railway and "foot" passengers. The Tunnel will transform the lot of the rail passenger, with through rail traffic providing Eurotunnel with around 35 to 40 per cent of its gross revenue — the rest coming from the road vehicle ferry trains and subsidiary activities which may include duty-free sales.

Revenue from through rail traffic was one of the bones of contention between the national railways and Eurotunnel, who were due to sign an agreement covering the level of tolls for through rail services in the Tunnel by the end of April 1987. The deadline was postponed on several occasions, and there was much posturing and brinkmanship on both sides. A deal was finally concluded in a late-night negotiating session at Heathrow airport on May 11.

The agreement covers the tolls that the railways will pay to Eurotunnel for every passenger and tonne of freight carried, but under the British Government's ideological stricture that no public money should be spent on the Tunnel, British Rail could not guarantee a particular level of payment to Eurotunnel because that would be tantamount to using public money, British Rail (BR) being a nationalised industry. With Eurotunnel

Model of shuttle trains at the Tunnel portal on the edge of Castle Hill. Here, passengers from France will have their first sight of the UK, with the large terminal at Cheriton. British passengers travelling in the other direction pass first through a cut-and-cover tunnel across Holywell Coombe and, less than 35 minutes later, they exit into the even larger terminal at Fréthun.

anxious to satisfy its bankers, and the railways keen to ensure that they are not paying over the odds, dispute was perhaps inevitable.

One of the other issues at stake was the speed of trains through the Tunnel. Each car shuttle train will take 27 minutes from portal to portal, this being a standard "path". Trains going faster than this will demand more than one path, because otherwise they would catch up with the previous train; slower trains would also demand more than one path as they would be caught up by the following train. Whereas Eurotunnel initially agreed on a maximum speed of 160 kilometres per hour, there is considerable pressure for the through passenger trains to run at 200 kilometres per hour. This is because BR and SNCF have their eyes on an hourly high-speed London – Paris service that may well prove a traffic bonanza similar to the Japanese Shinkansen (new trunk line) or the TGV service between Paris and southeast France. The British are more conservative in their traffic forecasts than the French, who envisage that nearly 16 million passengers will make use of the through-trains in 1993, rising to over 20 million after ten years. Eurotunnel estimates that car and coach traffic on its rolling motorway trains will amount to 13.8 million passengers in 1993 and 15.7 million in 2003.

The shorter the journey time between the two capitals, the better the prospects for traffic. Thus 200 kilometres per hour operation through the Tunnel might just keep the hand on the clock from reaching the 3 hour mark for a London – Paris trip. BR and SNCF are officially quoting a journey time of 3 hours 15 minutes, with London – Brussels taking 2 hours 55 minutes, but there is scope for whittling these timings down if the railways' plans come to fruition.

Centre-stage is a high-speed railway to be built between Paris and Brussels in time for opening of the Tunnel. This will form the first part of an ambitious project known as TGV-Nord which envisages construction of a new line from Paris to Brussels, Cologne and Amsterdam.

TGV-Nord will form a crucial link in the emerging high-speed network in Europe that is already centred on Paris. From the railways' point of view, plugging Britain into the European high-speed rail network would be the most significant effect of the Tunnel.

High-speed network

Europe's high-speed network was spearheaded by the French Paris – Lyons TGV line (called Paris – Sud-Est to emphasise its role in serving the whole of southeast France). Its spectacular success led to construction of a second TGV route, from Paris to western France, known as TGV-Atlantique. The first tranche is due to open in 1989, with trains running at a breathtaking 300 kilometres per hour — the same speed that is envisaged for TGV-Nord. TGV-Atlantique (and later TGV-Nord) will be by far the fastest railway in the world, although trains may one day go faster still — a specially prepared TGV peaked at the remarkable speed of 380.4 kilometres per hour in February 1981; this is the world rail speed record for a passenger-carrying train.

It is not only the French who are building high-speed railways to handle the inter-city traffic of the 21st century. The first section of a new high-speed railway running for 100 kilometres between Mannheim and Stuttgart in West Germany was commissioned in May 1987, and this will be followed in 1988 by the opening of the first stage of a second line running 326 kilometres from Hannover to Würzburg down the eastern flank of the country. Regular services at 250 kilometres per hour will be launched in 1990–91 using a fleet of Intercity-Express trains now being

Lorry and car loading trials on a full-scale shuttle model were undertaken on March 20, 1987. The Secretary of State for Transport, John Moore, was present.

developed from a five-car experimental train which reached a West German record speed of 345 kilometres per hour in November 1986.

In Italy a super-railway between Rome and Florence called the *Direttissima* (very direct) is nearing completion. Spain too is planning to build new high-speed railways, while tiny Switzerland will towards the end of 1987 hold a referendum on proposals for construction of four relatively short sections of new 200 kilometres an hour routes.

The origins of TGV-Nord can be traced back to the last Tunnel project — the proposals had to be scrapped when the British pulled out of the scheme in 1975. Revival of the scheme in the early 1980s quickly attracted the attention of the Dutch, and a branch was drawn on the maps to connect with the new line through Schiphol Airport that opened between Leiden and Amsterdam in 1981. The TGV-Nord idea was at first enthusiastically backed by the transport ministers of France, Belgium and West Germany, but the momentum slowed when Monsieur Chirac's right-wing government took office in France in 1986. Monsieur Pierre Mehaignerie, whose ministerial portfolio includes transport, did not display the same enthusiasm as his Communist predecessor Monsieur Charles Fiterman who had campaigned for what he called "a renaissance of rail".

The revival of the Tunnel had a profound effect on the gestating TGV-Nord scheme, as it quickly became unthinkable to build the high-speed line without a link to Britain through the Tunnel. The Tunnel increased the rate of return to such an extent that the French minister said towards the end of 1986 that without the Tunnel the TGV-Nord project could not go ahead. Already the French Government is contemplating an official report on the route of the section of TGV-Nord that lies in French territory. So popular is the French TGV that towns near the alignment of TGV-Nord such as Amiens are crying out for it to be diverted so that they can enjoy their own service.

No new line to London

The £400 million BR is planning to spend on cross-Channel trains and alterations to lines and stations in the London area must meet the Government's strict criteria for railway investment. It does not include anything as ambitious as a new high-speed line.

BR insists that it cannot build a high-speed line between Cheriton and London, as it was just such a project which castrated the last Channel Tunnel scheme in 1975, when the Labour Government said that the cost of the new rail link was too high. BR had failed to appreciate the economic necessities of the day and it had also run into a tremendous barrage of opposition to the new line which had forced it to bury much of the route in Tunnel, making the cost escalate out of control. This time it is determined not to make the same mistake and is limiting improvements on the line from Cheriton to London to a modest upgrading that will permit trains to run no faster than 160 kilometres per hour.

From Cheriton where the Tunnel trains will surface on the UK side, the route to London lies through Ashford, where a new international station is to be built. This will provide a major interchange between international and domestic services, and BR's Network Southeast sector is already examining the opportunities for running through-services from south coast towns such as Brighton. This will give added impetus to plans to electrify the 43 kilometre line across Romney Marsh from Hastings and Rye to Ashford.

From Ashford, Tunnel trains bound for London will normally be routed through Tonbridge and Sevenoaks, but BR is providing for alternative routes, for example through Maidstone East. This is already used by boat trains when the primary route is closed for track maintenance work. From Sevenoaks the route lies through Bromley South to Herne Hill and Brixton in the south London suburbs. At Stewart's Lane, a new section of track will be built to give access to the main line into Waterloo, which will be the London terminus for international services. Five new platforms will be constructed on the north side of the station where the Windsor line platforms are at present located. New platforms for domestic commuter services will be shoe-horned in elsewhere, for example on the site of the carriage road. Surprisingly, there will be fewer

The shuttle model in Ashford, Kent. This is not necessarily the final design as there are at least another four years left in which to plan improvements to deal with every eventuality.

trains using Waterloo after the Tunnel opens in 1993 than in 1971 when there was more commuter traffic.

The international trains will be serviced at a depot to be built at North Pole next to the main line from Paddington to Reading. To gain access to the depot, trains will run from Waterloo over a short spur to be reinstated between Clapham Junction and Queenstown Road so that they can reach the West London line. This would be substantially rebuilt and electrified, partly at 25 kV and partly at 750 V dc, as it will also be the route for international passenger and freight services heading for destinations north of London.

The fully air-conditioned high-speed trains that will work the London – Paris and London – Brussels expresses will blend British and French technology and design. A performance specification has been drawn up by the three railways concerned (BR, SNCF and Belgian National Railways, SNCB). They will pack in high performance traction equipment able to accelerate them up to 300 kilometres per hour on the new TGV-Nord line. A 17-car formation 400 metres long will seat around 750 passengers. Technically, the trains will have to use proven technology, but the demands on that are considerable. They will have to accept power at 25 kV 50 Hz for the high-speed line in France and for routes north of London, 750 V dc third rail for the Southern Region section, 1.5 kV dc for lines in the suburbs of Paris, 3 kV dc for running in Belgium off the high-speed line, and possibly also 15 kV 16⅔ Hz for operating through to Cologne in West Germany.

A range of catering will be provided, with restaurant and buffet facilities. There will also have to be accommodation for immigration and customs staff. While car and coach passengers will undergo customs and immigration checks at the shuttle train terminals, passengers on the through-trains will probably have their passports and luggage inspected on board the trains while travelling. This is common practice all over Western Europe, and even when crossing the Iron Curtain, passengers are not made to disembark. After initial opposition, the British customs and immigration authorities relented in favour of on-train inspection, and clear Government approval seems virtually certain, at least for journeys beyond London.

Trains beyond London

A favourable decision was crucial to BR's plans for through-trains north of London, because if passengers were made to get off the train at Ashford or Kensington Olympia, for example, the time penalty and the inconvenience of moving baggage would have destroyed any chance of competing with other modes. Criticism about unfair competition with the ferries was hardly the point as the British immigration service sometimes inspects passengers' passports on board the ferries. On the assumption that on-train inspection would prove possible BR planned a wide range of through passenger services beyond London. These will serve both the East Coast route to York, Newcastle and Edinburgh (which will be fully electrified by 1991 with London – Edinburgh journeys pared to just 4 hours), and the West Coast route to Birmingham, Liverpool, Manchester and Glasgow.

BR believes its chances of breaking into the long-distance package tour market, for example from Manchester to Venice, are quite low, as the airlines are already well established with low prices. None the less, an enterprising group of West German travel agents purchased a fleet of 33 special coaches to form trains known as the TUI Holiday Express in 1979. These are comfortable up-market couchette cars, with only four berths per compartment, lockers and tables; each train includes a club car with shop, bar and telephone where passengers can congregrate for drinks and snacks en route. Here they can also watch films or slide presentations about their holiday resorts or reminisce on the way home. TUI Holiday Expresses operate over relatively long distances, for example from north Germany into Yugoslavia, and there could well be scope for a similar operation through the Tunnel with destinations such as the south of France, Brittany, Switzerland and perhaps Austria.

Sleeping-car trains will run over medium distances such as Birmingham to Paris and Manchester to Brussels, and there may even be scope for restoring a London – Paris sleeper service, this being the only way of ensuring a comfortable overnight journey with an arrival time in Paris that allows a business appointment before 10.00. Even the earliest

Ashford station area.

flight from London does not permit an arrival in the centre of Paris much before 10.00, effectively ensuring that a British businessman has to spend a night in Paris.

Freight traffic

Less glamorous, but of major significance, are the prospects for freight traffic through the Tunnel. Container traffic in particular is expected to be a major growth area, but the closure of eight out of 22 Freightliner container depots in April 1987 did not help to bolster confidence. This left Glasgow as the only Scottish Freightliner terminal. Despite this short-sighted move — the result of Government pressure to reduce BR's call on the taxpayer — BR's Channel Tunnel Director Mr Malcolm Southgate confidently predicted that up to half the freight traffic through the Tunnel will be carried in containers. Glasgow is one of several sites being examined for a new deep-sea container terminal in Britain that could compete with Rotterdam by cutting more than 24 hours off the shipping leg of the journey for containers bound for the continent.

Around 75 per cent of the freight traffic would originate or be destined for areas away from London and south east England, and BR and Eurotunnel launched a regional information campaign in March 1987 with a special exhibition train touring 14 centres.

It is the long distances which will give rail the opportunity to compete more effectively with lorries, and Transport Minister David Mitchell predicted that hundreds of lorries a day will be taken off the roads, switching to rail. This does, however, depend on the willingness of companies to invest in rail handling facilities such as private sidings — ever since the Beeching days of railway closures, the number of private sidings on the BR network has declined. Inland freight clearance depots would also encourage rail traffic through the Tunnel, and BR is in touch with development agencies in a bid to ensure that important new centres are rail connected.

BR's plans for freight traffic include purchase of around 20 electric locomotives for hauling trains through to France; like the passenger trains, these will have to be able to accept power at 25 kV 50 Hz overhead and at 750 V dc for the part of the journey on Southern Region tracks. The £400 million investment includes funds to electrify the 32 km between Redhill and Tonbridge at 750 V dc, as this route will by 1993 be cleared for handling 2.59 metre (8ft 6in) containers.

Other freight train facilities will include an eight-track yard at Dollands Moor close to the Cheriton road vehicle shuttle train terminal. The yard will serve primarily for safety inspections before trains transit the Tunnel and for attaching an extra locomotive — every train will have to have a loco at each end to meet the requirement for driving it in either direction in the event of an emergency. Passing loops are to be constructed at Lenham, Tonbridge and one or two other locations.

Particularly exciting is the possibility of the high-speed freight train. This is a field which has until recently been neglected by the West European railways. Just what can be done is demonstrated by the French TGV postal services which whisk mail between the capital and Lyons every day in specially built TGVs carrying modular containers. They are more efficient than the small air freighters which they replaced.

SNCF introduced its first 160 kilometre per hour freight trains in the summer of 1987, but the genuinely high-speed freight train running at passenger train speeds of 200 kilometres per hour or more opens up unexploited opportunities. Here there is a chance to compete for small high-value packages at premium rates, a business at present dominated by air freight, courier services or dedicated fleets of lorries and vans. Even the West Germans are looking tentatively at plans for a high-speed freight train, the ICE-G (for Güterzug). Were this kind of high-speed freight service to run through the Tunnel, a package of medical supplies handed in at Birmingham, for example, could be in a Brussels hospital 4 hours later.

Another concept which has great potential is the RoadRailer and its derivatives. The idea is that a standard lorry trailer is fitted with railway wheels and is thus truly bi-modal, so avoiding the problems of transhipment. It runs in the the rail mode for the trunk haul, and becomes a standard lorry from the railhead to the consignee. The technique was tried in Britain in the 1960s but ran into opposition from the trade unions. Experiments with RoadRailers took place in the USA in the 1970s, but only in the last year or two has the idea caught on. Its significance in the context of the Tunnel is that it is probably the only technique that would enable lorries to go by rail in Britain because even lorries loaded on the West German "piggyback" wagons with tiny wheels infringe BR's restrictive loading gauge. Several companies are working on this and similar designs, but BR has yet to be convinced.

The Tunnel will bring substantial amounts of freight traffic to BR during the construction phase. Concrete lining segments fabricated on the Isle of Grain will be moved to the Shakespeare Cliff construction site where there is already a reception siding left over from the last project. Apart from that, materials for construction of the shuttle train terminal will also be moved by rail. On the French side tenders have already been called for construction of an 8-kilometre branch line from Calais to the worksite at Sangatte.

Apart from the new high-speed line, investment in railways in France will include substantial remodelling of Paris Gare du Nord where SNCF wants 11 platforms able to accept the 400-metre-long Channel Tunnel trains; reconstructing the station was costed at Fr2.5 billion, which even SNCF rejected as too expensive. The planners are hard at work trying to reduce the cost. Investment is also intended nearer the Tunnel portal — SNCF will electrify the 62-kilometre line from Hazebrouck to Calais for freight traffic.

Computer control

As freight and high-speed passenger trains approach the Tunnel from Ashford and Lille they will be monitored by Eurotunnel's computers controlling trains in the Tunnel area. These computers will allocate priority between different types of train and set the points for through-trains or car shuttle trains departing from or arriving at the terminals at Cheriton and Fréthun. Although the through-trains will be operated by the national railways, they will be under the control of Eurotunnel for their passage through the Tunnel.

Tunnel train drivers will have signal aspects displayed in the cab so that they can "see" some distance ahead, but as traffic builds up there will be a good case for fully automatic driving to ensure that headways between trains are kept to a minimum and the capacity of the Tunnel maximised. Automatic driving of trains is now a well established technique, with the Lille metro operating driverless trains since 1983 and the Vancouver Skytrain since early 1986. Line D of the Lyons metro will be driverless when it opens in 1990, and even the Paris metro is seriously contemplating the same technique. In London the Docklands Light Railway opened

North Pole area on the West London line was brought to the select committees' notice by a schoolboy. Rolling stock depot site viewed from across the main line. New shrubs are to be planted by the flats in the background to screen the sidings.

by the Queen on 30 July 1987 will be driverless too, although there will be a member of staff on board each train to provide information and check tickets.

On opening of the Tunnel in 1993, up to 20 trains an hour will pass through in each direction. Because of the speed differential between different types of train, there may only be 12 services an hour at certain times of the day, perhaps with four high-speed through-trains and eight rolling motorway car shuttle trains.

The concept of a rolling motorway is not new, but it is unfamiliar to the British to the extent that many people have no clear idea of how the Tunnel will work for motorists.

Rolling motorways through the Alps

Cars have been ferried on trains through the Alps since January 1924. In that year the Swiss Federal Railways launched a car-carrier service through the 15-kilometre Gotthard Tunnel on the important international transit route from Germany to Italy through the heart of Switzerland. The Gotthard and its sister trans-Alpine rail tunnels quickly assumed major importance for motorists, as they offered the possibility of all-year-round journeys through the Alps where the mountain passes were frequently closed in winter — just as they are today.

The 19.80-kilometre Simplon Tunnel between Brig and Iselle in Italy began a car-carrier service in May 1933, and by 1938 over 10,000 cars a year were being transported in this way. The 14.6-kilometre Lötschberg Tunnel on the Bern–Lötschberg–Simplon Railway (BLS) has also carried cars on trains for over 50 years, while a similar service has existed for many years in Austria through the Tauern Tunnel between Bockstein and Mallnitz.

The great boom in car traffic began in the 1960s, and in 1963 the three Swiss tunnels together carried a total of 465,442 cars. Traffic built up steadily, and in 1972 well over half a million cars transited through the Gotthard. It was, however, the Lötschberg which attracted most traffic, and in 1981 the BLS moved over 700,000 cars between Kandersteg and Goppenstein, with a record 9,260 cars, vans, small lorries and coaches on 28 February. By 1986 the Lötschberg Tunnel alone had ferried 10 million cars under the mountains. In the same year the BLS reduced its fares on the car-ferry trains, resulting in traffic burgeoning upwards; 882,000 cars — with an average of three people riding in each car — were carried safely through the double-track Lötschberg Tunnel.

Most BLS car-carriers shuttle to and fro between Kandersteg at the north end and Goppenstein at the south end of the Lötschberg Tunnel, but some motorists elect to use the services that save the steep drive down into the Valais through the Lonza gorge and stay in their cars on the train as far as Brig; here the BLS meets the Simplon main line from Paris to Milan which continues through the Simplon Tunnel into Italy. A number of BLS car-carrier trains run on through the Simplon Tunnel to Iselle in Italy. This journey from Kandersteg to Iselle through the two tunnels takes nearly one hour - almost twice as long as it will take motorists to transit the Channel Tunnel in 1993.

In 1980 the Gotthard car-carrier trains were withdrawn following opening of the parallel motorway tunnel. There was no point in Swiss Federal Railways (SBB) attempting to compete for the car traffic, as no charge was levied for use of the motorway tunnel. Then, in 1982, a car-carrier service was launched through the 15.4-kilometre Furka Tunnel of the metre gauge Furka–Oberalp Railway between Oberwald and Realp. Although car-carrying operations are on a smaller scale than the other Swiss tunnels, the route provides a valuable east–west link in the south of Switzerland, immune from the ravages of winter weather. Before the Furka Tunnel was completed, even the railway had to be closed in winter where it passed over the mountains near the Rhône glacier that lends its name to the famous *Glacier Express* which runs between St Moritz and Zermatt.

To begin with, cars were carried through the Swiss tunnels on completely open, flat wagons, but special trains were later built that offered roofs as a rudimentary protection against falling ice. The same design remains in use today. To date, no fewer than 25 million cars have been ferried under the Alps, 16 million of them in Switzerland alone over the last 16 years.

Far from being an outmoded form of transport, car ferry trains have been chosen as the primary mode for another trans-Alpine link. In 1986 the Swiss Government gave the go-ahead for the 19.1-kilometre Vereina Tunnel, in the southeast corner of Switzerland between Klosters and Lavin on the metre gauge Rhaetian Railway; as with the Channel Tunnel and the other Alpine tunnels, car-carriers will share track with ordinary passenger and freight trains. Starting next year, construction of the Vereina Tunnel will be completed two years after the Channel Tunnel — progress will be much slower because the Vereina has to be carved out of hard rock rather than through the relatively soft Chalk Marl under the Channel. Anyone using the car-carrier services in Switzerland cannot fail to be struck by the extreme simplicity —even crudeness — of the arrangements. It is a straightforward drive-on, drive-off operation, the only touch of sophistication being the opportunity for passengers to tune their car radios to a special tunnel radio wavelength.

The Channel Tunnel shuttles

In contrast, the Channel Tunnel car-carrier trains — or shuttles, as Eurotunnel prefers to call them — will be fully enclosed and considerably more sophisticated than their Swiss cousins. There will be three types of shuttle train: double-deckers for ordinary cars, single-deckers for coaches and caravans, and single-deckers for lorries.

The double-deck shuttles will be the largest passenger-carrying trains built to date. Amtrak's twin-deck Superliners on long-distance trains in the USA are one metre longer but not quite as high, while a fleet of double-deck commuter trains in Belgium has coaches which are 2.5 metres longer, as is the latest type of long-distance single-deck passenger coach in the Soviet Union. The shuttles will be about 4 metres wide compared with trains in the USSR which are 3.4 metres wide. Shuttle train design was not finalised at the time of writing, but as they will not be built until 1990-91 for delivery towards the end of 1992, this was not a problem. Design was in the hands of the Anglo-French Transmanche-Link (TML) joint venture , which has the design-and-build contract for the Tunnel (Chapter 3). The fundamental design principles, however, are already decided, and an overall picture of the shuttle trains can be

Variations on a theme. Most fixed link ideas envisaged rail transport facilities but not necessarily as a first priority, and some placed the trains in an immersed tube.

assembled. Full-sized mock-ups of the shuttles have already been con-structed — they were inspected by Secretary of State for Transport John Moore in Ashford, Kent, on 20 March 1987 — and further models will be built in due course to assess details as design progresses. One problem will be the carrying out of running trials, as the shuttles will be too large for both the British and French rail networks. There is some possibility of testing space-frame prototypes in the USA or Canada, where they would not infringe the loading gauge. Were full-sized prototypes to be construc-ted, they could even be put into regular service for a short period in North America — which would be a sure way of locating any problems.

Ordinary motorists will generally use the double-deck tourist shuttles, ten of which will be built for the start of services in 1993. The end wagon of each 15-wagon set will be specially designed with wide side doors for loading and unloading on both decks. The car shuttles will not be short of space. Internal width of a wagon will be 3.75 metres, so that if 1.75 metres is assumed to be the average width of a car, less than half the internal width is taken up. So even with a car door open at 45 degrees there is still more than 0.5 metres left for people to walk past. Car drivers will have to apply their handbrakes after parking in their allotted positions on the shuttle, the floor of which will be constructed of a special non-slip surface, so that even during an emergency brake application the cars will not move. None the less, Eurotunnel has been studying whether some form of chocking might be necessary; it may be that the single-deck shuttles carrying coaches and lorries will need chocks.

On the double-deck shuttles, stairs will be located in every third wagon and in the loading and unloading wagons at the end of each rake; lavatories will be provided under the stairs. There will be less space next to the stairs so cars will not be parked here. Cars with disabled people will be identified when loading so that they can be located near exits on the lower deck. Eurotunnel plans to have a refreshment service available, but has yet to decide what form this will take. Even the possibility of stewardesses serving passengers in their cars has been canvassed, although the number of crew that this would require would be in marked contrast to the Swiss car-carrier trains which are driver-only operated.

Eurotunnel currently envisages that six staff (including the driver) would man each shuttle: a double-deck shuttle will have space for 126 cars, compared with up to 75 cars (and one member of staff) on the most modern Lötschberg car-ferry trains. English and French speaking staff on each shuttle will be highly trained personnel who rotate on a range of duties. This will ensure familiarity with all aspects of the operation — which may be important in an emergency — and it also relieves tedium.

The second type of tourist shuttle will be single-deck, and this is designed for coaches and motorists with caravans. There will be accommodation for about 30 cars, each with a caravan, or their equivalent in coaches.

Eurotunnel plans to operate a single-deck and a double-deck shuttle in tandem to give plenty of flexibility. Each of these giant trains will thus have the ability to carry the equivalent of nearly 200 cars — about two-thirds of the capacity of a cross-Channel car ferry and more than three times that of the largest hovercraft which ply the Channel between Dover, Calais and Boulogne. At opening, shuttle services will run every 12 minutes in peak periods. Because of the frequency of service, no advance booking will be necessary, and average waiting time should not exceed 15 minutes, which will include the frontier formalities. Journey time through the Tunnel will be 27 minutes, with a start-to-stop schedule from terminal to terminal of 35 minutes.

Lorries separated

The third type of shuttle will be for lorries — providing a much welcome separation between passengers and freight which are crammed together on the present ferry services. A lorry shuttle formed of 25 wagons, each 21 metres long, will be able to carry 25 fully-laden lorries. There will be special loading wagons for the lorry shuttle trains; these loading wagons may be open with drop-down ramps on each side so that lorries can manoeuvre to line up before driving into the shuttle. An alternative design has loading wagons with sliding telescopic hoods similar to wagons on many railways used for transporting steel coils that must be shielded from the weather. Inside the lorry shuttles there will be restricted clearances, although no worse than on the ferries at present. Lorry shuttles will operate about every 20 to 30 minutes.

As space is limited at the British terminal at Cheriton, lorries will pass through customs at a clearance centre located inland at Ashford; from where there will be direct access to the M20 leading to the shuttle terminal. Carrying lorries on trains is not a new technique. It has been common practice in Western Europe and the USA for many years. In France, SNCF uses so-called "Kangourou" wagons with pockets for the trailer's road wheels to nest in. In West Germany, Austria and Switzerland special low-floor flat wagons with tiny wheels (known as piggyback wagons) have been developed. In the USA many railways use skeletal wagons in which the lorries' wheels sit in wells just above the running surface of the rails to give maximum height, and a similar type is being developed in France.

Each shuttle train will be worked by a pair of electric locomotives, one at each end. A combined power rating of around 10,000 kW is envisaged, so they will be among the most powerful locos in Europe. Even if one loco were to break down, its twin would be capable of hauling the train and the dead loco out of the Tunnel by itself. Eurotunnel will have a fleet of 40 of these locos, together with four diesels for use on maintenance trains and in emergencies. Maximum speed of the shuttle trains will be 160 kilometres per hour, but this will usually only be on the downgrades soon after entering the Tunnel. The ride quality should be quite high, with modern suspension on the bogies ensuring a quiet and smooth trip. The specification for the shuttle trains will demand an interior noise level of less than 80dB(A). Electronic controls on the locomotives will ensure smooth acceleration and braking — no comparison at all with the pitching and tossing of the ferries which sometimes results in road vehicles being damaged in rough weather. Nor will there be the vibration which makes hovercraft travel such an unpleasant experience.

Before loading commences — a process which should take about ten minutes - each shuttle will be cleaned and inspected. Before the train departs it will be "scavenged" of foul air to avoid a build-up of exhaust fumes. Prior to driving aboard at one of the ten loading and unloading

Swiss shuttles carry cars through the mountains.

platforms in the terminal (provision will be made at both the Cheriton and Fréthun terminals for six more platforms to be added), shuttle passengers will have completed all frontier formalities. On arriving at the terminal, each car driver will obtain an "intelligent" ticket from a toll booth. The ticket will be encoded with information on vehicle type and category.

Customs and immigration will be arranged on a "free exit" basis, with both English and French checks carried out before the shuttle is boarded. After obtaining a ticket, someone travelling to France from Cheriton would next pass through the British emigration and customs control before reaching a duty free area. On leaving this, he would undergo French immigration and customs checks before reaching the shuttle boarding zone where the intelligent ticket will be inserted at a barrier. A computer will read and analyse the information on the ticket and assign a departure platform. This will be printed on the ticket, which is returned to the customer, who will then drive over bridges and ramps to the platform.

Safety in the Tunnel

Opponents of the Tunnel have latched on to the question of the safety of
passengers travelling in their cars on the trains, citing in particular the
danger of fire. The flames were fanned by a scaremongering video
launched in 1986 by Sealink British Ferries, the cross-Channel ferry
company set up by container magnate Mr James Sherwood when he
bought BR's ferry services for £66 million in 1984. This blatantly anti-
Tunnel video was widely cited by the British press, who had apparently
forgotten that in 1985 Sherwood had proposed a drive-through Tunnel in
which the chances of a fire breaking out would be considerably higher
than in a shuttle train where the cars have their engines switched off.

The video attacked the idea of passengers travelling in their cars on the
grounds that they are not allowed to do so on the ferries. This apparent
logic is a nonsense, because the interior of what is effectively a ship's hold
is a dramatically different environment from the interior of a train. The
movement of a ship makes the car decks a relatively dangerous area
because of the possibility of vehicles shifting, something which is most
unlikely on a train and impossible if some form of chocks is used. But the
fire question prompted Labour MP for Deptford, the late John Silkin, to
say that the Tunnel could become the longest crematorium in the world, a
piece of emotional wordmongering which delighted Flexilink, the ferry
opposition group, and his near neighbours in Deans Yard, Westminster
in their campaign against the Tunnel.

Not once in the 63 years that motor cars have been carried through the
Alpine rail tunnels has there been a fire or other incident involving loss of
life on a car-carrier train. Only on one occasion in the early years of
operation did injuries occur when the couplings parted and a car fell
between the wagons; since then, cars have not been allowed to park over
the couplings. The only other incident was a small fire in the Simplon
Tunnel involving an Italian diesel railcar — not a car-carrier train — which
was caused by faulty maintenance the previous day; the train crew acted
swiftly and correctly so that no-one was injured or killed.

The spectre of catastrophe was none the less raised by Lord Dean of
Beswick in the House of Lords on 16 February 1987 when he asked Lord
Sanderson of Bowden if he recalled that "just after the war there was
possibly one of the biggest train disasters in history which occurred in
those Alpine tunnels? Over 400 people were suffocated when a train
broke down in a tunnel." The noble Lord appeared to have the wrong
mountains, for he was presumably referring to an incident which
occurred on March 2 1944 in the Apennines in southern Italy. A freight
train hauled by two 2-8-0 steam locomotives left Balvano for Potenza on
the Battipaglia – Metaponto line. Travelling slowly, as the engines were

struggling to lift the 47 wagons up the grade, the train entered the Armi Tunnel where it ground to a halt. Poisonous fumes in the smoke of the two engines rapidly filled the tunnel; they were particularly noxious because of incomplete combustion resulting from the poor quality of the coal. The engine crews must have died quite swiftly, and in normal circumstances they would have been the only victims. But it was wartime, and the goods train was being used by about 600 desperate people as a means of transport across the mountains. Nearly all died, apart from a lucky few who were at the rear of the train which did not actually enter the tunnel.

Remarkably, in the Swiss tunnels there is no communication between the cars on the trains and the train driver, who is however in radio contact with control staff outside the tunnel. Were a fire to occur, special fire fighting trains stationed at strategic points would be mobilised, but the Swiss have never had to send them into the tunnels — they have been more use helping to extinguish mountainside brush fires! On the Alpine trains a fire could theoretically sweep back from car to car with flames blown rapidly by the rush of air in the tunnel, but this particular hazard would not apply in the Channel Tunnel, as the Eurotunnel shuttles will be fully enclosed.

Eurotunnel has worked out a comprehensive fire prevention and emergency strategy. The philosophy is to minimise the chances of a fire occurring, to ensure rapid detection if it does, and then to put it out or contain it. If this is not possible, then the Tunnel will be evacuated, and Eurotunnel has to meet a statutory requirement of clearing the Tunnel of all passengers within 90 minutes. Eurotunnel is planning its fire defence strategy in close consultation with the Department of the Environment's Fire Research Station. All safety measures will have to be approved and monitored by an Anglo-French Safety Authority. This will be chaired on the British side by the Chief Inspecting Officer of Railways in the Department of Transport, who will have to assist him the Chief Highway Engineer in the Department of Transport, the Chief Fire Officer of Kent and a senior officer of the Health & Safety Executive with a matching group of French officials.

To prevent fires breaking out, passengers will be required to obey a number of regulations: motorists with caravans will not be allowed inside them to brew up tea, caravan calor gas taps must be turned off and smoking will be prohibited. Passengers will be made well aware of the potentially serious consequences of breaking these by-laws, which will be enforced by train staff. Incidentally, passengers are allowed to smoke on the trans-Alpine car-carrier trains.

The shuttle trains will be constructed to withstand fire for more than 30 minutes so that they can continue out of the Tunnel where it is easier to combat fire. The emergency alarms will not apply the brakes for the same reason, but will alert the train crew to initiate emergency procedures. Fire doors or curtains to prevent fire and smoke spreading from one vehicle to another will be provided, and small doors in these fire screens will permit passengers and staff to move from one vehicle to the next. Fire detection equipment and extinguishers will be fitted in each shuttle vehicle for use by passengers or staff in emergency. Should this fail to put out the fire, passengers will be evacuated to the next vehicle.

If a fire breaks out the primary intention will be to take the train out of the tunnel — it will never be more than about 15 minutes from the open air. If the fire — or any other disaster — is serious enough to require the train to be stopped, then the offending vehicle or set of vehicles will be uncoupled with passengers evacuated to the adjacent shuttle wagons. The rest of the train will then be removed from the tunnel. Any passengers unable to go with the train will be evacuated to the central service tunnel where the air pressure will be kept higher than in the running tunnels to prevent smoke being blown into it. At least two doors will be available on each side of a shuttle wagon for emergency escape. Once out of the shuttle, passengers will be able to move along the continuous walkway on the inner side of the running tunnels from which cross-passages lead to the service tunnel every 375 metres, so that there will be at least three passages within the length of an 800-metre shuttle train. All three tunnels will be permanently lit.

Once in the service tunnel, passengers will be evacuated by a train from the other tunnel or by a special transport system within the service tunnel. This may consist of a narrow gauge railway or rubber-tyred vehicles.

There will be a 250-millimetre diameter high-pressure water main fed from both ends located in the service tunnel. This will supply fire hydrants located at regular intervals in both running tunnels and spaced less than the length of a train apart. Other fire fighting equipment will be provided, possibly with special trains stationed at each end of the tunnel. Another line of defence is the ventilation system. The flow of fresh pressurised air can be increased at the coast to force it into a running tunnel at the rate of 700 cubic metres per second, to blow smoke away from a stranded train, for instance.

It is already clear that hazardous goods such as liquefied petroleum gas carried in rail tank wagons will not be permitted through the tunnels — explosive and corrosive chemicals will be carried on the ferries, as they are now. Incidentally, according to the Belgian Minister of the Environment Mrs Smet, there were 50 drums of toluene cyanate, 61 drums of cyanide and 200 packages of soluble lead - all highly toxic substances - on board Townsend Thoresen's ill-fated *Herald of Free Enterprise* which overturned in a matter of seconds at Zeebrugge on 6 March 1987, resulting in at least 196 people losing their lives.

That tragedy muted the campaign of the ferry companies against the Tunnel.Not that they have given up. I was inspecting a mock-up of the shuttle trains in Ashford on 20 March 1987, watching a lorry simulating the loading procedure. I was photographing the proceedings when a man on my left pointed to the gap between the wall of the shuttle wagon and the lorry and said "Now just imagine a woman trying to get out of there."

I was just about to point out that it would be extremely unlikely that a woman would be on the lorry shuttle unless she happened to be a lorry driver and that the space was in any case no less than between lorries on a car ferry, when a security man and a Eurotunnel official grabbed my elbows and attempted to hustle me away. "Right off the site, sir!" I protested, produced my credentials and received a grovelling apology. My companion, it turned out, had gatecrashed the site as a representative of the ferry companies and was endeavouring to stir up trouble.

Other people have suggested that the Tunnel will be a top target for terrorists. But blowing up trains is not nearly as effective as blowing up aircraft where disaster is practically certain. To destroy a long train by a bomb planted within it would require an immense and detectable quantity of explosives distributed along its length, whereas with an aircraft a relatively small single device would suffice. A bomb planted in the Tunnel aimed at derailing a train might be more effective, although even then the tunnel walls would keep the train upright (it is when railway coaches overturn that deaths and injuries are most serious because people are thrown around), but strict security precautions will guard against unauthorised people gaining access to the Tunnel — just as airports are protected. Cameras and other devices will monitor the tunnel mouths, although Eurotunnel is understandably reluctant to divulge exactly what security measures are planned.

Recent experience bears out the evidence that modern trains are not easily destroyed. Three bomb attacks have been carried out on the French TGV. The first was on 11 May 1981 before the new Paris - Lyons line was opened when one of the TGVs was running in ordinary service. An explosion took place in the luggage area in the third car; the emergency brakes were applied and the train halted from a speed of 150 kilometres per hour in a distance of 1500 metres. Four passengers were injured, but they were released from hospital on the same day. The fire-resistant design of materials and equipment on board the train prevented the spread of fire.

More serious was a bomb that exploded on a TGV travelling from Marseilles to Paris on 31 December 1983 and which was again travelling at 150 kilometres per hour. Three people were killed and 11 injured. Despite an explosion which SNCF described as "very large", the train did not derail, although three of the TGV's cars had to be repaired or rebuilt.

The most recent attack on a TGV took place on 17 March 1986 when a Paris – Lyons train was brought to a halt by emergency brakes coming on automatically as a result of an explosion damaging the brake pipes. Again the train was running at 150 kilometres per hour, and again it did not derail. Nor was there a serious fire thanks to the fire-retardant properties of the materials used in the train.

Before leaving the subject of safety, it should be remembered that one of the prime reasons for building the 54-kilometre Seikan railway tunnel between the islands of Honshu and Hokkaido in Japan — which when it opens in 1988 will be the world's longest rail tunnel, 4 kilometres longer than the Channel Tunnel - was the *Toyamaru* ferry disaster when 1,430 people died during a storm that wreaked havoc in the strait between the two islands.

Including the Zeebrugge tragedy, there have been at least four major incidents involving ferries in the Channel in the last five years. The Channel shipping lanes are among the busiest in the world, and three collisions occurred in fog in a single day on February 4 1987. Hovercraft are vulnerable too, as was shown when one of them collided with a breakwater in Dover in March 1985, killing four people. Their cross Channel journeys are frequently cancelled when the wind gets up, in winter months.

Safety is relative and never absolute. While the opponents of the Tunnel fulminate about crematoria they would do well to remember that people who drive to France will be taking a much greater risk by driving on French roads than sitting in their cars on a train in the Channel Tunnel. No fewer than 10,961 people were killed on French roads in 1986 alone, and a further 259,000 injured.

The emotional scare about rabies is equally irrational. To meet Government requirements, import and transit controls for animals will be in force just as they are at ports and airports, backed by dog patrols searching for smuggled pets. There is no reason to suppose that these measures will be less effective with the Tunnel — they have ensured that only 19 people have died of rabies in Britain since the Second World War.

To prevent animals moving into or colonising the Tunnel all entrances and drains will be fitted with mesh to keep out rodents. Small-mesh fences around the Tunnel boundaries will act as the first line of defence, and these will be sunk into the ground to prevent animals burrowing beneath them. Where the fence is broken, cattle grids or pits will form traps, and electrified mesh will be installed just outside the tunnel portals. There will be no food supply in the Tunnel as trains will be fully sealed so that no debris can be dropped. No animal is going to venture for 50 kilometres through a foodless concrete desert!

Like the supposed threat of military invasion, rabies is dug up by the Tunnel's opponents every time the issue is raised. Were rabies a major problem on the continent, then the British would need to be vaccinated against it every time they crossed the Channel, just as they need to take malaria pills when they visit Africa or certain parts of the Far East.

One other safety problem deserves to be aired. Eurotunnel is committed to examining the case for a drive-through tunnel under the terms of its concession. Before going too far down that road, it should bear in mind what is happening in the much shorter Gotthard motorway tunnel. Swiss engineers are having to reckon with the possibility that from 1990 onwards the ventilation equipment will not be able to cope with the build up of diesel fumes, leading to increasingly frequent closures of the tunnel every time the pollution level reached a predetermined threshold. That is one reason why plans to build a 50-kilometre railway "base" tunnel under the central Alpine chain are gaining momentum. Among the many advantages of an electrified rail tunnel under the Channel with a shuttle service for road vehicles is that the tunnel air will be almost as fresh as the sea air above.

References

[1] *Railway Gazette International*, May 1987, p.263.
[2] Behrend, G. and Buchanan, G. *Night Ferry*, Jersey Artists, 1985.
[3] *Railway Gazette International*, June 1987.
[4] Fischer, Dr J. Holiday coaches to penetrate the package market, *Railway Gazette International*, September 1979, p.906.
[5] Chaîne, Dr H. Lyon adopts unmanned trains, *Railway Gazette International*, May 1987, p.291.
[6] *SBB CFF FFS*, Publication of the Swiss Federal Railways, Berne; edited by Dr E. Schenker.
[7] Scheider, A. and Masé, A. *Railway Accidents of Great Britain and Europe*, (Translated from German by E. L. Dellow). David & Charles, 1970.
[8] *Le Monde*, March 10 1987.
[9] *Japanese Railway Engineering*, No. 100, December 1986.
[10] *La Vie du Rail*, No. 2082, February 19 1987.
[11] Sperren am Gotthard wegen Lastkraftwagen, *Tages-Anzeiger*, January 24 1987.

5

The Environmental Impact

John Ardill

In March 1986, Ms Penny Evans of the Kent Trust for Nature Conservation and Ms Trina Paskell of the Royal Society for Nature Conservation went to the *Guardian* newspaper in London, to explain their worries about the Channel Tunnel. They suggested ways that Eurotunnel and the UK Government might reduce the Tunnel's impact on the natural environment and also gave some ideas of "constructive" uses for the chalk spoil which will come pouring to the surface as the triple bores are driven under the Channel. These ranged from a dramatic landscape setting for the Cheriton terminal to the artificial recreation in some suitable spot of chalk downs, the rolling sheep walk country rich in flowers and insects which was once widespread in south and east England, but has now largely disappeared under the farmer's plough and pesticides.

The spoil, unfortunately, is not that amenable to creative use. The greyish, bread-loaf-sized hunks cut by the tunnelling machines will shake down into small powdery lumps or a putty-like slurry by the time it reaches the open air. The more it is moved about, the messier and more difficult to handle it will become. It will be like yoghurt, say the French, who plan to tip three million cubic metres in a dry valley near the Sangatte terminal, burying three old blockhouses, and another 250,000 cubic metres is to be placed in two disused quarries. Both places will be relandscaped. While much of the British spoil will be used in conjunction with colliery waste to make the foundations of the terminal and create a working platform at the base of Shakespeare Cliff, much will be left over. How to dispose of it in an environmentally acceptable manner has been one of the most vexed questions of the planning and legislative phase of the project. The current plan, strongly opposed by the UK Government's statutory environmental advisers as well as local and national conservation lobbyists, is to keep it where it arises at Shakespeare Cliff. The House of Commons Select Committee on the Channel Tunnel Bill was sufficiently impressed by arguments that it will do more damage there than elsewhere to ask the Government and Eurotunnel to look again at alternative sites which had already been intensively examined and agonised over [1]. The matter was raised again in the House of Lords Select Committee but the conclusion was that Shakespeare Cliff was the most suitable site. A proviso was added to ensure that spoil resulting from any future fixed link construction could not be dumped at the same location.

The early preoccupation of the conservationists with the spoil not only highlighted what the Commons' Committee agreed was a "very serious question". It also defined the scope of effective action by environmental lobbyists once the UK Parliament had decided that the project should go ahead. That decision could not be challenged: all subsequent effort had to be put into minimising the environmental damage and maximising the opportunities for compensatory environmental improvement.

Like most works of transport infrastructure, the Tunnel will have both environmental costs and environmental benefits. Cross-Channel traffic is expected to double over the next 20 years and without a fixed link large additional areas of land would be needed at the ferry ports for car parking, terminal building and for new berths. Existing main roads in Kent would probably cope with the extra demand for the next 15 years, given already planned improvements, but they, and the M25 ring which gathers traffic up from the national motorway network, would come under increasing pressure. Rail freight services through the Tunnel, estimated by British Rail to take 1500 lorries a day off the roads, will relieve some of that pressure and reduce the noise and pollution of road transport. Millions of air passengers may also switch to less noisy, less polluting and more energy-efficient rail transport, and that could also help reduce the future land needs of airports, although at the cost of funnelling passengers through city centre stations. Fewer ships at sea would decrease accident rates, spillages of oil from tankers or chemicals from ferries, even unanticipated chemical cargoes as were found on beaches after the *Herald of Free Enterprise* tragedy in March 1987.

The diversion of traffic from roads, however, will not be as great as might have been with a simple rail tunnel without shuttle facilities, hence the muted enthusiasm for the scheme of some pro-rail lobbyists. Nor will the Tunnel necessarily prevent an environmentally damaging extension of Ramsgate's ferry port in Thanet. To the extent that it helps generate economic growth, the Tunnel may also provide a stimulus to, and resources for, environmental improvement. But that depends on where and how consequential development is undertaken. If it is directed to currently rundown areas, the environmental benefit could be considerable: if it is confined to areas of Kent and the South East already under development pressure, the damage will be incomparably greater.

Moreover the Tunnel is being promoted not to divert traffic to a more environmentally-acceptable mode but to encourage competition between modes and stimulate economic growth. While it will take a share of an increasing amount of traffic, it will also generate more traffic. It is now widely accepted that building roads does not so much ease congestion as encourage more journeys; and in the sense that the shuttle will be a "rolling road" it will probably have this effect. As both a spur, to and symbol of, economic growth the Tunnel is unattractive to those who believe the future of humanity and our planet depends on a less intensive use of material resources. Even to those who see potential good in it, this is not a project in which a careful balancing of direct environmental costs and benefits can be the final test, as it may, for instance, over proposals for tidal barrages on the Severn and Mersey estuaries. Judgement must be made between the economic benefits and the environmental costs. If the Tunnel is built some, but not all, of those environmental costs will be exacted long before the economic benefits can be reaped. Others will flow in the train of whatever economic growth the Tunnel helps generate. It is not beyond the bounds of possibility that, along with other changes now taking place, it will help redraw the economic, social and environmental map of Britain.

What then are the environmental effects? They are of three kinds: land irretrievably lost or severely damaged, including its landscape, nature conservation and amenity values, by the construction of the Tunnel and its associated works; disturbance during the construction and operation of the system, including visual intrusion, noise, dust and other pollutants; and the effects of consequential developments locally and in the country at large.

Samphire, seaweed and fen

The south east corner of Kent is the butt end of a long hump of chalk upland, confusingly known as the North Downs, which was severed from mainland Europe only some 8000 years ago. Its people are said to be the most Continental of all Britons, and some fancy this is due not only to proximity and the ferry trade, but also to a lingering folk memory of the time when it was possible to walk across to the mainland. Proximity, geology and climate make it the only toehold in Britain of some European plant and insect species which, with others more characteristic of Britain, make up a particularly rich wildlife resource of major scientific interest. Kent as a whole has rather more of its surface under some form of wildlife protection than the average for English counties. The white cliffs, where the undulating surface of the Downs meets the sea, are a vital ingredient in every Englishman's mental and emotional image of his homeland. Combined with the south-facing escarpment, serrated by deep coombes, which runs west from Folkestone and overlooks the Weald, they constitute a landscape treasured as one of the finest in southern England. When you add to the wildlife and landscape values, the existing pattern of settlements, roads and railways, which is also conditioned by land forms, this becomes a difficult place in which to insert the essential works of the Tunnel system. It is not surprising therefore to find that the elements of the scheme will not be contained in a single site as in France but strung out from Dover where the tunnels cross the coast, through the shuttle terminal and railway inspection sidings at Cheriton and Dollands Moor, west of Folkestone, and into Ashford's inland clearance depot for road freight and an international passenger station.

At the foot of Dover's Shakespeare Cliff is a four-hectare platform of material blasted from the cliffs in 1840 when the coastal railway was built and known, because of undersea mining in the 1890s, as the Old Colliery site. It is from here that some of the previous attempts to tunnel under the Channel were made, and the site is littered with the debris of abandoned workings, including segments of prefabricated tunnel lining from the 1974 venture. Their removal by the Government in advance of the new project will be an environmental bonus point, the significance of which was dramatically illustrated by the discovery that a quantity of the highly toxic and persistent chemical polychlorinated biphenyl (PCB) had been spilt on the ground during earlier salvage operations. Several hundred tonnes of contaminated soil will have to be excavated, under carefully-controlled conditions, and taken away for disposal, probably by high temperature incineration. The British end of Eurotunnel will be dug from here, and a permanent service complex will be established, comprising ventilation fans and a cooling plant, emergency services, a water store for firefighting, and other buildings. To accommodate these works it is planned to extend the platform to a minimum of 14 hectares using 1.9 million cubic metres of chalk from the excavations. In the absence of an alternative disposal site, a further 1.85 million cubic metres of spoil will be added, creating within a seawall a platform reaching 200 to 300 metres out to sea and stretching for a about 1.5 kilometres, level with the railway. The House of Lords Select Committee added a proviso that no more spoil from a future fixed link should be dumped on this site. On the cliff top, near the village of Aycliffe, will be a construction site with temporary buildings and a car park, and a permanent ventilation shaft. Somewhere near by will be a temporary camp for construction workers.

The cliffs rise at a slope to a height of some 130 metres. "Half way down", said Edgar in *King Lear* "hangs one that gathers samphire; dreadful trade!" Samphire is a cousin of Hemlock and the more familiar roadside Cow Parsley. In Shakespeare's time its fleshy stalks and leaves were a delicacy, cooked like asparagus. It still grows on the cliffs, along with Sea Cabbage and the Yellow Horned Poppy, all common coastal plants, and other more rare species like the Sea Health which is usually found in salt marshes. Half the cliff face is covered by vegetation, and this is a key to its particular importance to conservation scientists. The cliffs between Dover and Folkestone, largely unprotected by sea defences, are naturally eroding, and subject to occasional landslides. Erosion and exposure to seaspray give rise to five different communities of plants, from the most salt-tolerant near high-water mark to chalk grassland species at the cliff top. There is a rich insect community too, including four extremely rare breeding species of moth, joined in season by migrant moths and butterflies.

The whole area from the top of the cliff to low-water mark between Dover and Folkestone has been designated since 1951 as the Folkestone Warren Site of Special Scientific Interest (SSSI). It is a grade 1 site, that is, of international importance, and its scientific interest, both biological and geological, is complex and varied. Botanists have been studying its plants for the last 400 years, and it is one of the birthplaces of the science of geology. The chalk cliffs are a national "standard reference"" for rock of this kind and their eroded slope makes them more accessible for study than the vertical cliffs nearby, while the erosion process itself is important for the study of geological forms. Outcrops of Lower Greensand and Gault Clay are also important geologically. The waters of the Channel, close to the boundary between two oceanic regions, combine with the rock types to create a particularly important marine environment. On beds of Gault Clay, some of which will be covered by the working platform, all of which may disappear if the platform is extended, are colonies of small seaweeds which, the Nature Conservancy Council (NCC) says, are probably not found in this precise form anywhere else. The NCC, which has a statutory duty to identify and conserve SSSI and to advise the Government and the public on their protection, believes that the Tunnel works will destroy, or seriously damage much of the interest of the site. The effects will be both direct, as with the burial of the Gault beds and other exposed rock surfaces, and indirect. The presence of the platform may stabilise the cliffs, halting the process of "rotational slip" and covering the whole surface with vegetation. This would reduce the variety and interest of plant species and make the rock surfaces inaccessible to study. Beyond the platform the erosion process could be accelerated. On these grounds the NCC is advising against the scheme. But it is also collaborating with other organisations in technical studies to determine the best design of the seawall and platform to minimise the impact on the cliffs and regulate the deposition and erosion of material along the shore. Care will also be taken to guard against pollution of the waters with chalk spoil and subsequently with Tunnel and site drainage water contaminated with oil, biocides and other chemicals which could damage the marine biology and the inshore commercial fisheries. The clifftop site is outside the SSSI but is nevertheless botanically valuable as an example of unimproved chalk grassland containing a good population of Early Spider Orchid. The Kent Trust for Nature Conservation is

The area to be changed most will be around the hamlets of Newington, Peene and Frogholt. These sheepish local residents live happily alongside the M20 motorway and do not seem unduly perturbed at changes taking place around them.

pressing for the site to be restored after use to as near as possible the condition of the surrounding land.

The coastal strip lies within the Kent Downs Area of Outstanding Natural Beauty, a designation made by the Countryside Commission and confirmed by the Government in 1968, which signifies that the landscape is of the highest quality, and of national importance. It is further defined by the Commission as a Heritage Coast because of its "significance in both landscape and recreational terms as one of the finest stretches of undeveloped coast in England and Wales". From Dover to Folkestone and along the top of the scarp runs the Downs Way, one of the Commission's 13 long-distance footpaths. Like the NCC, the Commission wants the platform limited to an essential working area so that it obtrudes as little as possible on the view from the cliff tops. Otherwise, on a slightly hazy day. the Commission's regional officer Ian Mitchell told the Commons' Committee "the sea views will become views of an artificial platform." Dover District Council is prepared to have the surplus spoil at Shakespeare Cliff if nowhere else is found but wants all the available space after completion of the works laid out for public recreation. The conservation bodies would prefer it to be kept primarily as a nature reserve. The Commission's wish to have the platform landscaped to ameliorate its appearance conflicts with the NCC's priority of maintaining the scientific intererest.

From Shakespeare Cliff the Tunnel will run parallel to the shore to Sugarloaf Hill, near Folkestone. It will cross the Holywell Coombe in a cut-and-cover section — in fact a shallow cut and a raised cover forming a ridge across the coombe — before disappearing underground again at Castle Hill. A new route for the A20, which is associated with the Tunnel and included in the Bill, goes through this area also. Holywell Coombe is part of the Folkestone to Etchinghill Escarpment SSSI. Among its botanical treasures is the Late Spider Orchid, a Mediterranean-region flower. The population in Britain is thought to be no more than 100 plants and it has legal protection as a rare and endangered species. Its presence at Holywell, says the NCC's regional officer Dr Timothy Bines, is "a remarkable and historic" example of how an exotic plant can take advantage of a rare combination of factors, which include the warm southfacing slopes, the shelter of the surrounding land and the rich calcareous soils [3]. There is a second endangered species, the Bedstraw Broomrape, and other rarities like the Woolly Thistle and the Wild Cabbage. The coombe also has an area of fen where springs have formed wet marshy ground, unusual in a chalkland area and therefore of particular scientific interest

What the environmentalists' photographs do not show are the road and light industrial warehousing adjacent to the site, making it less idyllic than usually depicted.

Within the coombe are two small deposits of silt, clays and rocks laid down during the late Ice Age and containing important fossils of plants, beetles and snails, in a combination not found anywhere else. According to Dr Keith Duff, the NCC's head of geology, they are of fundamental importance both for pure research and for the industrial application of geology in, for instance, the identification of oil-bearing strata [4]. The

NCC is concerned about the possible destruction of these features and about the effect of dust from the tunnelling works on the varied insect life in the SSSI. Thick dust, it suspects, could interfere with their breeding and feeding. This effect will be monitored during the construction period and may yield useful information although the chances of preventing any damage seem remote.

In a package of measures to minimise damage overall, Eurotunnel has offered to drop its original plan to bore the landward section of the tunnels from Holywell Coombe and will work uphill from Shakespeare Cliff. Once a service tunnel is cut, more than 1 million cubic metres of chalk will be brought through the Coombe to make the foundations at Cheriton. Under pressure from the conservation bodies, and after site investigations, it has adopted as its preferred route a more southerly alignment of the tunnels than originally planned, to avoid damaging the fen and one of the two vital geological deposits. It has also offered £100,000 for a rescue dig of the second deposit. At the time of writing it had not, however, given a firm commitment to the realignment. The southerly alignment would destroy one of the known locations of the aptly named Late Spider Orchid. Elsewhere within the site, locations of this and other rare plants will be protected from the construction work by fencing.

The inland port

From Castle Hill the Tunnel emerges at the Cheriton terminal site, 140 hectares of mainly agricultural land which has been safeguarded by the planning authorities for this purpose since the previous Tunnel proposals. The site, shaped like an elongated pear, lies below the escarpment, bounded on the south by the M20 which separates it from Cheriton and on the west by the villages of Newington, Peene and Frogholt. To the west across the M20, on a narrower strip of land beside the railway line, will be the Dollands Moor freight inspection sidings. The intercontinental railway line will cross the M20 on a viaduct. Dollands Moor reaches into the Kent Downs AONB and the Folkestone to Etchinghill SSSI. Fifteen houses will have to be demolished, and land will be taken from five farms. Village cricket and rugby grounds will be displaced and a number of footpaths closed. The only significant feature of nature conservation interest that will be totally destroyed is five hectares of ancient semi-natural woodland called Biggins Wood. The wood has a rich ground cover of flowers and a good range of mosses and lichens but it is not uniquely valuable. Attempts will be made to transfer some of the plant species to a new landscape site. Seven known archaeological sites, including a Romano-British farmstead and an ancient field system, will be destroyed: others are likely to be discovered and will need recording before they are lost.

The main impact of this huge "inland port" will be on its rural surroundings. The Cheriton and Dollands Moor sites together stretch for some 4.5 kilometres, and with the adjacent motorway will cover an area almost as big as Dover Harbour, around 2.5 hectares. The terminal site alone is as big as Hyde Park, in London. Newington, Peene and Frogholt, which cluster around the western end of the terminal, are attractive old farming villages now, like many others, occupied more by the retired, commuters and other relative newcomers. To the south of the M20 are the Cheriton area of Folkestone, army camps, and the village of Saltwood. About 65 village properties lie within 500 metres of the site and are liable to suffer excessive noise during construction, and a disturbing level of noise during operations. While embankments and landscape planting will help shield them from the sight and noise towards the end of the construction period and when the terminal is in operation, they will replace an open view across fields with a dark, dense mat of vegetation

Mr Robin Grove-White, director of the protest group, the Council for the Protection of Rural England. He is not against the Tunnel in principle but wants the bad effects minimised. CPRE accept that improved rail links can cut down on the more polluting road traffic, and would certainly accept the Eurotunnel scheme as the lesser of fixed link evils.

only 100 metres from the nearest house. The rail viaduct across the M20 will be less than 150 metres from the 12th century Church and visible from virtually everywhere in Newington — "a very substantial and significant eyesore" the Parish Council said [5]. The council argued for the railway to be taken under the M20. This was ruled out for operational reasons but Eurotunnel has managed to lower the designed height by about 2.5 metres, creating great technical problems along the route. Residents of Cheriton will be able to see the terminal platforms, gantries and buildings; those of Saltwood will see and hear the freight train movements in Dollands Moor.

No amount of landscaping will soften the alien intrusion into the view from North Downs Way, where ramblers seeking only the beauty of the countryside will be liable to rub shoulders with others wanting to watch the terminal at work. A viewing platform and car park planned on the escarpment to accommodate sightseers caused a great deal of concern to the House of Lords Select Committee. Nature conservationists are worried about the impact of too many feet on the SSI chalk grassland. A hang gliding centre — which has no security of tenure — will lose what its proprietor Mr Damon Robinson says is the best training site in the UK. A model aeroplane club will also be displaced.

Under the original plans, access roads from the M20 to the terminal would sweep around the north of Frogholt and Newington and enter the terminal through a cut-and-cover tunnel between Newington and Peene, leaving the two villages surrounded and isolated. The route cuts into the AONB; very close to Asholt Wood, a Grade 1 SSSI, and could damage its scientific interest by interfering with the drainage. After lengthy negotiations and debate before the House of Commons Select Committee a new access route was agreed, close to the continental main line and the M20, containing the access roads within the existing corridor. The alternative was put forward by Shepway District Council, not only to reduce the environmental impact but also to make it easier for cars leaving the terminal to turn into Folkestone, thus increasing the town's chances of gaining trade from the Tunnel. The alternative would however require the demolition of four or five houses in a conservation area, but was accepted by Parliament following recommendation by the House of Lords Select Committee.

The Tunnel proposals have already had a severe impact on villages. Local petitioners told the Select Committees of increased incidence of ill health, including respiratory and blood pressure problems, probably due to anxiety about the scheme. Mr Barry Pattinson said he had been ill for three days when told that his home, Mill House, Newington, would be demolished for the rerouted access road: "the Doctor diagnosed Channel

Tunnel syndrome" [6]. Eurotunnel has offered to buy the houses of all who wish to leave, and by March 1987 had taken over 25 of the 83 properties registered in the scheme. At that time Mr Percival Derham, chairman of the Parish Council, told the Committee that Newington had an air of depression, with every other house empty, and that its character was probably changed for ever [7].

The Dollands Moor facility was strongly opposed during the Bill proceedings by local authorities, amenity bodies and local residents. Kent County Council, Newington and Saltwood Parish Councils, the Council for the Protection of Rural England [8] and many local residents argued that the sidings should be accommodated within the terminal; others that they should be sighted at Ashford. The terminal site is considerably more confined than the one on the French side. Shepway Council, again with an eye to economic advantage, argued that if it must have the sidings at Dollands Moor it should have the international passenger station there too, as British Rail had intended up to 1985. The House of Commons Select Committee was persuaded by Eurotunnel and British Rail not to change the arrangements, for reasons of operational necessity and safety because, it was argued, the environmental impact would be worse at Ashford.

The local cricket and rugby club will lose its playing field to the terminal at Cheriton. Although an alternative site has been offered, the players will probably lose one season as the ground has to be prepared first.

Ashford District Council petitioned against losing the international station, which British Rail will build, together with long-stay parking for lots of cars, as an extension of the existing station. Ashford is the only town in Kent planned for significant urban and industrial expansion, and the council welcomes the growth-generating potential of the station and the customs-clearance depot for freight lorries which will be built to the south east of the town in an area zoned for industrial and warehouse development. The depot will be connected to the M20 by a new south orbital road, part of which is already built. Authority to build the rest of it, and to realign the A2070 connecting the new road to the town centre and railway station, is included in the Bill. The depot will occupy 7 hectares, and landscaping a further 4 hectares, of a 50 hectare site Eurotunnel seeks to acquire on the south side of the road. Amenity bodies, fearful that this will mark the start of a string of Tunnel-related industrial sites along the M20 to Folkestone, unsuccessfully urged Parliament to reduce the landtake and locate the depot on the Ashford side of the new road.

Spoiling the environment

The Tunnel is a massive engineering project. Millions of cubic metres of spoil must be removed from the earth, and millions of tonnes of materials brought to the construction sites, over a five year period. All this will involve considerable noise, dirt, fumes and other pollutants, a heavy volume of traffic, and the temporary occupation of sites while the work is in progress. Increased levels of road and rail traffic, and the concentration of vehicles and activity in the freight clearance depot, the rail sidings and terminal, will extend these environmental burdens into the future. Detailed arrangements are required for controlling and monitoring these operations and ameliorating their environmental impacts.

   Disposing of the surplus spoil from the tunnels has been one of the most intractable problems of the whole project. Few people other than civil engineers are convinced that adding it to the Shakespeare Cliff platform is a particularly good solution. For Eurotunnel it has the merit of being the cheapest and most convenient option. The difficulty has been to find an environmentally-better solution. Using any more distant site, however safe and suitable, as a final resting place, would incur the cost of pollution and noise from transporting up to 28,000 tonnes a day of messy material by rail, road or conveyor — possibly all three — throughout the tunnelling period. Detailed studies by consultants started with a list of 71 potential sites produced as the most likely alternative Lappel Bank, by the Isle of Sheppey, where an area could be reclaimed for port or industrial use. This was supported by Kent County Council and the local Swale District Council who invited a planning application. But using Lappel Bank would mean sending 21 train loads of spoil a day across Kent, and although British Rail was willing to take on the job, Eurotunnel put the additional cost at £40million. Its consultants, Environmental Resources Ltd, advised that Lappel Bank could not be considered a reliable option

Stone Farm, Ashford Road, Newington. A listed building to be demolished to accommodate the joint southern access route to the terminal...and its death sentence attached to a post.

for more than half the surplus. The Government argued that there would be technical and environmental problems, including possible damage to fisheries from an accidental spillage of chalk, and also voiced doubts about the long-term stability and therefore industrial usefulness of the disposal site. The CPRE, unconvinced by Eurotunnel's estimate of the additional cost, has continued to argue for this option, disregarding the disturbance to residents whose homes the spoil trains would pass.

Eurotunnel's plan to take 1.4 million tonnes of colliery shale from Betteshanger pit in East Kent — making a significant improvement in the local environment — for the foundations of the Cheriton terminal raised considerable alarm over the prospect of up to 300 lorries a day traversing the narrow lane out of the coalfield. There are similar fears about the impact of construction traffic on minor roads and low grade major roads throughout Kent. Eurotunnel promised the House of Commons Committee that all the shale would be carried by rail, an undertaking subsequently amended to allow the last stage of of the journey to be made by lorry on the M20 because a railhead could not be provided early enough at Cheriton. It also promised its best endeavours — subject to striking a balance between "practicalities, environmental concerns and economics" — to move as much bulk material as possible by rail. Eurotunnel told the

Commons that up to three quarters of the estimated 9 million plus tonnes of bulk material could be sent by rail, the rest going by road on routes authorised by Kent County Council. But it refused to meet the request of amenity bodies for a formal undertaking that at least two thirds would be consigned to rail. Fears remained that circumstances dictate that a larger proportion goes by road and there were doubts about the County Council's ability to enforce its restrictions by the rigour demanded by the Commons' Committee; these concerns have eased after the Lords' Committee urged similar restrictions to those suggested in the other house.

The construction of the Tunnel system will interfere with the existing hydrology of the sites, while both construction and operation phases will generate additional wastewater. At Cheriton and Holywell Coombe, efforts will be made to divert water arising from springs along the foot of the North Downs escarpment to the Seabrook and Pent streams so that it does not add to the volume of rainwater accumulating on the site. Water on site during construction phase will be polluted with chalk and clay and with oil from machinery, and surface water from the terminal will also be contaminated. Eurotunnel's plan to build a lagoon on the Seabrook Stream to hold and dilute the surface run off met with strong opposition from the NCC, which has the stream listed as an SSSI. The main scientific interest lies in the botanical richness of its associated fens which support 66 species of cranefly — an exceptional number. The fens would be drowned by the lagoon and the water quality down stream could be altered, possibly endangering the natural ecology and a commercial trout farm. Under strong pressure from the NCC and other conservationists, Eurotunnel is considering replacing the lagoon by a drainage pipeline to the sea at Folkestone, but it has made no firm commitment to this option. Following a recommendation in the House of Lords Select Committee report, this is the most likely course of action.

Existing environmental conditions in all the areas involved were surveyed, and forecasts of likely changes made, in an Environmental Impact Assessment (EIA) submitted by Eurotunnel as part of its bid for the fixed link franchise [9]. This is an established procedure for major development proposals which was specifically required by the terms of the competition. The report by Eurotunnel's consultants, Environmental Resources Ltd, is the basis both for proposals to ameliorate the impacts of the scheme and for continuing scientific studies. The EIA says the existing air quality around the Shakespeare Cliff, Cheriton and Ashford sites is typical of unpolluted semi-rural and rural areas. It forecasts that construction traffic is not likely to bring a significant increase in exhaust pollution levels given the current amount of traffic. Dust from the construction activities poses more of a problem. Without effective controls, concentrations are likely to exceed EEC limits, causing health problems to residents by exacerbating respiratory problems, reducing visability, and settling to a troublesome degree on plants and surfaces, including water in the reservoir at Castle Hill. Newington, Peene and Frogholt face the biggest problems, especially if the biggest route for the terminal access roads had been selected. The Dollands Moor works may cause significant dust problems at Saltwood, particularly in dry windy weather.

With the Tunnel in operation, concentration of vehicle exhaust gases around Cheriton are expected to be within the criteria set by relevant EEC and British standards. Even under conditions of severe traffic congestion and poor atmospheric dispersion, carbon monoxide concentrations are likely to be less than half the Department of Transport's hourly peak standard. The only other place of concern is the emergency ventilation shaft at the top of Shakespeare Cliff where smoke with a high carbon monoxide and hydrocarbon content from a Tunnel fire could cause a significant impact on the settlement of Aycliffe if the wind was blowing that way.

Noise levels in Newington, Peene, Frogholt and on the edge of Cheriton will, during the early months of construction in particular, exceed the levels usually considered acceptable for British construction sites, and protective measures will be needed, the EIA says. Peak noise levels during operation of the terminal may cause some problems in the three villages, even with site landscaping. At the Ashford clearance depot, night-time operation could affect six properties in Sevington.

Careful control of on-site operations, together with shielding embankments and tree planting, will help alleviate the noise and dust problems. In addition Eurotunnel has set up a number of compensation schemes for those most likely to be affected by its site during construction and operation. It has agreed to buy within ten years of 20 January 1986 any house in Newington, Peene and Frogholt, and any of a limited number in Sevington affected by the inland clearance depot. It will also contribute to removal expenses. The scheme [10] provides for independent valuation based on local market prices but disregarding the existence to the Tunnel project, and for arbitration in the event of a dispute. The valuation formula is similar to that applied under legislation to properties which may be compulsorily purchased to enable the project to go ahead. The second scheme [11] is a procedure for settling without recourse to the courts, small claims — up to £2000 — for damage to property arising from the works. Damage to fishermen's equipment near Shakespeare Cliff is included. Thirdly, Eurotunnel has agreed to provide double glazing in houses and in a school likely to be affected by noise from the terminal and the Ashford depot. This is an adaptation of the statutory provisions for noise insulation in respect of road schemes, but Eurotunnel has committed itself to a rather more generous noise threshold for triggering the scheme.

The House of Commons Select Committee welcomed the schemes but rejected a plea by Shepway District Council that they should be given statutory force. The Government has agreed however to appoint an administrator to act as a medium for complaints about any aspect of the Tunnel [12]. The Committee turned down a plea that Eurotunnel should be required to set up a capital fund to help deal with any unforeseen environmental or social effects of the Tunnel, accepting the Government's argument that there was no precedent for a such a move.

Safeguards and assurances

The statutory system of town and country planning, and other regulatory Acts, provides means for controlling civil engineering works and land used, including adverse environmental impacts. While the Channel Tunnel Bill itself gives the necessary planning permission for the scheduled works to be carried out, it leaves control over the construction

Landslipped chalk above the clay at Folkestone Warren, with vegetation special to the geology beneath. The Martello Tower has been converted into a house.

process, the detailed layout and design of buildings, and the operation of the facilities, to the local planning authorities [13]. It specifies that the construction must be carried out in accordance with a "scheme of operations" agreed between Eurotunnel and the authorities. This covers, in general terms, the sites for obtaining, storing and sorting minerals and aggregates and the methods and routes for transporting them; the hours of work on the construction sites; measures for suppressing dust and noise; the use of site lighting; and the siting of construction camps. The Bill does not give the authorities carte blanche. They are only allowed to alter Eurotunnel's submitted plans in the interests of controlling the depletion of mineral resources, maintaining the free flow of traffic, preserving amenity and sites of archaelogical and historic interests, and securing nature conservation. Any modifications required by the planning authorities must be "reasonably capable" of being carried out.

The planning authorities must also be satisfied with detailed plans and specifications for the erection, alteration or extension of buildings, car parks and noise screens, lighting installations, access to public roads and the construction of the Shakespeare Cliff platform and seawell. Again, they can only alter the submitted plans in the interests of amenity, or because a particular structure could reasonably be placed somewhere else within the scheduled area. In the case of the platform, which was brought within control at Kent County Council's request, they can act in the interests of nature conservation. In the face of fears that spoil from the French half of the Tunnel might be brought to the Kent end, the Commons' Committee wrote into the Bill a limit of 3.75 million cubic metres on the amount that may be deposited at Shakespeare Cliff. The figure refers to the uncut volume of chalk: once cut it will bulk up to some 5 million cubic metres. The authorities must be satisfied that noise suppression and landscaping measures have been carried out before the Tunnel and its facilities can be commissioned. In all these matters they must consult the Countryside Commission, the Nature Conservancy Council or the Historic Buildings and Monuments Commission when any of their interests are affected.

The Bill replicates normal planning procedures by allowing the Environment Secretary to "call in" any of the detailed plans and take the decision himself. This gives the statutory agencies, and the public, scope to seek Government intervention if they feel that not enough account is being taken of their views. But the Minister will have to be persuaded that he should act, and his freedom to determine the issue is circumscribed in the same way as the local authorities. In turn, Eurotunnel can appeal to the Secretary of State against the planning authority's decision. The Bill gives the local planning authorities control over the restoration by Eurotunnel and British Rail of construction sites when the work is finished and obliges the Government to restore sites to their original, or some other appropriate, use if the project is abandoned.

The planning provisions of the Bill are supplemented by a Memorandum of Agreement between Eurotunnel and the local planning authorities [14] which has the backing of (the Government and) Parliament. The authorities undertake to decide planning applications as quickly as possible, and promise not to impose any "unusually stringent" standards. Eurotunnel and its contractors promise to use their best endeavours to mitigate adverse environmental impacts "consistent with the need to maintain the programme and project viability". The Memorandum also makes it clear that controls over aggregates are not intended to restrict Eurotunnel's commercial freedom to get supplies from within or outside Kent, and that the planning authorities will not seek to impose movements by rail when this would involve an unreasonable cost penalty. The authorities accept that motorways, trunk roads and certain other A and B roads will be used for hauling bulk materials. Restrictions over working hours, and measures to keep down noise, dust and mud are spelt out.

In a further Memorandum of Agreement with the Nature Conservancy Council, Countryside Commission, Historic Buildings and Monuments Commission and the Southern Water Authority [15], which covers some of the same ground, Eurotunnel again promises that it "will adopt measures from the outset to mitigate any adverse environmental impact...and will further conservation interests." It also promises to monitor environmental impacts, carry out careful strict management measures, and cooperate with statutory bodies and local planning authorities. The Memorandum specifies a long list of site management and pollution control measures, the fencing-off of sensitive areas in and around the work sites, agreement on the excavation and recording of archaeological sites consultation with expert bodies, and numerous research and monitoring programmes during both the construction and operation of the system.

The Bill frees Eurotunnel from sections of the 1981 Wildlife and Countryside Act prohibiting operations likely to damage flora, fauna or features of SSSI. As originally drafted, it would have lifted the regulations in respect of any works carried out by Eurotunnel in connection with the system. Under pressure from Kent County Council among others, the Commons' Committee amended this to apply only to the sites of the specifically authorised works. In spite of the protective legislation, which is far from absolute, and the NCC's view that the SSSI network is the minimum necessary to support viable populations of most species, damage to SSSI by development in general and transport infrastructure in particular is not uncommon. A report published in 1986 by Friends of the

Earth [16], drawing attention to the damage likely to be caused by the Tunnel, listed 110 SSSI which had been or were threatened with damage by road schemes. A 1980 sample survey by the NCC found that 15 per cent of SSSI had been affected by loss or damage. This is not to suggest that any such damage is readily acceptable. As we have seen at Shakespeare Cliff and Holywell Coombe, many SSSI features are virtually irreplaceable and their loss is a matter of absolute not relative cost.

There is no specific legislation to protect areas designated as being of high landscape value, but the 1968 Countryside Act requires Ministers, Government departments and other public bodies to "have regard to the desirability of conserving the natural beauty and amenity of the coun-

Canal barge on the peaceful waters of the Nord–Pas de Calais region in France, with the cathedral of Notre Dame à Amiens in the background. Environmental groups are strangely quiet this time around, but provided strong protest at the last Chunnelling attempt in 1974.

tryside." Once again, intrusive development is not uncommon. The very nature of the upland areas most commonly designated as AONB or National Parks makes them liable to mineral workings and military training ranges, for example. Fine landscapes too have their particular characteristics which make each area unique. Any intrusive development represents a loss in that particular area, and a diminution of the total quantum of fine countryside. It also tends to be seen as a symbolic attack upon the concept of conservation, and a potential precedent for future incursions.

Setting aside the matter of initial planning consent, the Bill thus ensures that the project will be carried out under similar public control to any other building or engineering scheme. Moreover, Eurotunnel has gone to some great length, by adapting its plans and giving assurances and undertakings, to meet the objections of statutory agencies, voluntary bodies and the public at large. This has been acknowledged by representatives of the NCC and the Kent Trust for Nature Conservation who found Eurotunnel sympathetic and quick to grasp the point of their concerns. The Newington Parish Council chairman Mr Derham has also admitted finding Eurotunnel reasonable and sympathetic to deal with. The CPRE, veteran of many major public enquiries and Hybrid Bill hearings, is less impressed. It was critical, throughout the Select Committee hearings, of Eurotunnel's failure to produce detailed layouts and plans for buildings and landscaping, details which, it says, would normally be available at a public inquiry or Bill hearing. Mr John Popham, Parliamentary Agent for the CPRE and other amenity bodies, maintains that no major environmental concessions were granted during the committee hearings. Even winning the right to amenity body participation in the local authority consideration of reserved planning matters took a great deal of pressure, he says. Unlike the normal planning system, there is provision in the Bill for disagreements on planning issues to be referred to arbitration. The amenity bodies fear that the planning authorities will be unable to maintain the control they are allowed if Eurotunnel, finding itself under pressure of time or finance, sought to evade its undertakings. Only by getting undertakings embodied in the Bill can they be made to stick in any circumstances, the amenity bodies and other interested parties believe. While the major petitioners managed to convince the Select Committees to do this in some cases, they failed on many others. Some issues of which petitioners sought binding agreements were still in negotiation when the committees concluded their business. There remains the question of whether undertakings made before Parliament but not embodied in the Bill can be enforced. Mr Peter Boydell QC, counsel for Shepway District Council, reminded the Commons Committee that no one had ever established what happened in the case of breached undertakings — although it had been suggested that the culprit could, as in centuries past, be sent to the Tower of London!

Kent roads old and new

The construction and operation of the Tunnel will have a big impact on the road system of Kent, much of which is already heavily used and in places needs improvement to cope with existing or expected levels of demand. Kent County Council estimates that without the Tunnel, traffic to the Channel port would at least double by the year 2003 [17]. The principal routes, the A2–M2 and the M20–A20, have with scheduled improvements enough capacity to cope with this increase. The Tunnel should divert some of this potential growth to the railways but will also generate additional road traffic. It should certainly divert some of the traffic now using the more northerly A2–M2 corridor to the M20–A20 which the authorities have long agreed should become the main route for cross-Channel travel. The main concern in the short and medium terms lies therefore with the secondary routes and minor roads which may face a heavy burden both of construction traffic and of subsequent cross-Channel traffic with origins and destinations within Kent or to the West.

The package of Tunnel-related proposals includes a number of road schemes. Some, like the terminal access roads and to a lesser extent the Ashford schemes, are essentially part of the Tunnel system and are included in the Bill. Others, like already planned improvements to the M20, are being dealt with concurrently by the Department of Transport as the trunk road authority. Improvement of the A20 between Folkestone and Dover, which is included in the Bill, is a special case and will be dealt with separately below. In addition, local authorities and communities are anxious to see other roads improved, either to cope with traffic changes generated by the Tunnel or as a counterbalance for areas likely to suffer economically as a result of the Tunnel. Many of the problems exist quite independently of the Tunnel project and Parliament witnessed scores of petitioners seeking assurances that their pet schemes will be completed before the Tunnel opens: literally trying to "get in on the Act".

The M20 will be the main feeder road for the Tunnel — indeed its existence and its connection to the M25 London orbital route and thus to the national motorway network is an obvious rationale for having a fixed link capable of taking road traffic, albeit by shuttle. Construction of the missing link between Maidstone and Ashford was already in the DTp's programme but work is being advanced for completion by the end of 1989. The Department has commissioned studies on widening the Maidstone bypass section in time for the Tunnel's opening and although it is assuming one additional lane in each direction, it has promised to consider sympathetically Kent County Council's argument for four-lane carriageways.

The Department has also announced improvements on the A259 route running west of Folkestone to New Romney and Hastings for south coast traffic. The eastern end of Folkestone through Sandgate and Hythe will be downrated from a trunk road and an alternative trunk route will be provided by an improved A261 to the A20. Bypasses of New Romney, St Mary's Bay and Dymchurch, which had been put in abeyance, have been restored to the construction programme. They will involve an incursion into Romney Marsh SSSI. An eastern bypass of Hastings is being studied. Under pressure from the local authorities, the Department has promised additional funds for Tunnel-related road schemes which are the responsibility of the County Council so that other priority schemes in its programme do not suffer. Already extra grants have been provided for the Ashford South Orbital and the A2070 diversion to the Town Centre, included in the Bill, and for a bypass of Hamstreet further south on the A2070. Another A2070 bypass at Kingsnorth will be considered for the 1988 grant.

Other schemes put forward by the local authorities include the A256 from Sandwich to Dover, the A258 Deal to Dover, the A260 north from Folkestone to the A2 and the B2068 from Canterbury to the A20. All of

A new bridge will be built alongside this one at Ashford station in Kent and the timber yard on one side will fall prey to a compulsory purchase order.

these, the authorities say, are likely to carry heavier traffic generated by building and servicing the Tunnel as well as from Tunnel users. Improvements will in some cases also help access to industrial sites which might benefit from the Tunnel. The authorities have also argued strongly for improvements to the A299 Thanet Way, the main access to the Margate–Broadstairs–Ramsgate area, to help counter the economic pull of the Tunnel. Although the Tunnel could result in lower, rather than higher, traffic levels on the route, the Department has promised to consider the case sympathetically, while pointing out that the additional help for the routes directly related to the Tunnel will allow the County to keep the Thanet Way high on its own priority list.

Individual road schemes reflect in microcosm the mix of environmental costs and benefits seen in the Tunnel system itself. Where they take traffic away from built-up areas or off narrow lanes and other essentially local routes they benefit adjacent properties and communities. But new routes and widening schemes also mean taking more land, with a potential loss of building and archaeological sites, and of some loss of inherent wildlife and landscape values even if no sites of particular importance are involved. In the case of the A20 improvement, the landscape and nature conservation impacts are of the highest order.

The A20 is the cuckoo in the nest: a scheme of dubious relevance to the Tunnel which came to be the most contentious element in the Tunnel Bill. Certainly Dover District Council, which is counting on the improvement to help the town's harbour compete with the Tunnel, told the Commons that it regarded the scheme as the single most important issue in the Bill. That opinion is shared by the Countryside Commission and the amenity bodies, but for entirely the opposite reasons. The scheme is not so much directly related to the Tunnel as inversely related to it. Without the Tunnel, traffic on this route is expected to fall in the short term, although a higher proportion of it is likely to be made up of heavy trucks. Why then was it decided to include this improvement in the Bill when the Thanet Way, for instance, cannot be promised even a place in Kent's enhanced expenditure programme? The initial answer, when the Bill was intro-duced, was that the improvement was needed to handle construction traffic. This quickly proved false, as it became apparent that the new route, running close to the cut-and-cover tunnel in Holywell Coombe, could not be built until that section of the Tunnel was completed. To all those opposed to the scheme the real answer is quite evident: the Government is taking the opportunity to salve Dover's wounded fortunes by slipping through without the customary public inquiry, a scheme which meets virtually none of its normal criteria for road funding or routeing.

The scheme had been under active preparation since the cancellation of the previous Tunnel project in 1974. It was shelved in 1984 pending a decision on the Channel Fixed Link and the then Transport Minister Mrs Lynda Chalker said it was unlikely to be built if the Link went ahead. Its inclusion in the Bill came therefore as something of a surprise. In fact, only half the route, from the M20 at Folkestone to Court Wood, is included. The eastern section was left out, the Commons was told, because of uncertainty over the route in Dover where either a factory or an ancient monument would have to be demolished. The DTp has promised local authorities that the western section should be completed by 1992 and the eastern, to be handled under normal highway procedures, by mid-1993.

The route included in the Bill is based, with some modifications, on one selected by the Department in 1981 as having most public support. It sweeps north from the end of the M20 to clear Folkestone and the village of Capel-le-Ferne on the existing A20, running through the AONB and the Folkestone and Etchinghill SSSI. The route cuts the escarpment of the Downs, crosses Holywell Coombe on a viaduct, and passes through Round Hill in a tunnel. From there it runs along the Alkham Valley, rising to higher ground on an embankment and joining the present A20 at Court Wood. The eastern extension is likely to run down the Aycliff Valley, pass between Shakespeare Cliff and the village of Aycliff, in Kent, and enter Dover along the existing road, demolishing houses and other buildings.

Both the Countryside Commission and the Campaign for the Protection of Rural England called expert witnesses before the Commons' and Lords' Committees to argue that the forecast levels of traffic did not justify construction of a new road under the Department's cost-benefit criteria. And that although the existing route was unsatisfactory, particularly through Capel-le-Ferne, the bypassing of a relatively small settlement would not normally have high priority in the national roads' programme. The Countryside Commission reminded the Committee of the Transport Secretary Mr Nicholas Ridley's statement in 1982, that road schemes affecting AONB "will be examined with particular care to ensure that a new road is needed and that the route has been chosen to do as little damage to the area as possible" [17]. The route was opposed, they point out, by the Department's Landscape Advisory Committee which described one part of it — subsequently somewhat modified — as disastrous and another as "particularly difficult on landscape ground". The Department replied that although an economic rate of return was an important test of whether a road should be built, it was not the only one. It pointed out that the existing route on Dover Hill was the steepest trunk road descent in the South East.

The opposition to the scheme's inclusion in the Bill went three steps beyond the actual damage the new route would do to the protected countryside. Firstly, the Hybrid Bill procedure meant that opponents could not in theory question the case for the scheme and the chosen route,

The Coach House and the Grange in Newington will both be demolished when the terminal's access route is constructed. One indignant resident mounted guard at his window with a shotgun for a while, showing his displeasure, forcing up the compensation stakes and sending journalists scurrying away in trepidation.

only the detail. In the event, the Commons' Committee gave them a good deal of leeway in this respect, encouraged by a message from the Department that if an alternative means of improving the route could be found, it should be brought before the Committee. Secondly, fixing the western section route in the Bill virtually determines the route for the eastern extension, again denying opponents, including local objectors, much scope for argument at the eventual public inquiry. The Countryside Commission's regional officer Ian Mitchell told the Committee that the eastern extension would do even greater damage to the environment since the Department's favoured alignment passed along the Heritage Coast "devastating an exquisite dry valley immediately behind the cliff top, completely destroying the tranquillity of the coastal scene and this most important section of the North Downs Way". The scheme, he added "would be more damaging to the countryside of East Kent than all of the Channel Tunnel works put together." [19].

Thirdly, it might set a precedent for dealing with environmentally-contentious road schemes by Parliamentary Bill to avoid the normal procedure, which can greatly prolong the time taken to get them built. The conservationists were only too aware of the frequent criticism of public inquiry delays voiced by Ridley, Transport Secretary when the Channel Bill was drafted and now Environment Secretary. It was Ridley, too, who had defeated them over a similarly contentious road, the Okehampton Bypass, getting it routed through the edge of Dartmoor, contrary to normal Government policy against major new roads in National Parks. The Commons' Committee was not persuaded to delete the scheme, although Labour member Nick Raynsford argued in his minority report that it should be dropped.

The Battle of Waterloo

British Rail's case for Waterloo as the Tunnel's international rail terminal in London rests on environmental as well as economic and operational grounds. The new platforms and other facilities needed will be contained mainly within the existing station, although the approach viaduct will have to be enlarged. Only four buildings will be lost, the most prominent of which, Northcote House, is used in part by Lambeth Council as a treatment centre for young offenders. Of the dozen locations considered, the two other serious contenders both fall down badly, British Rail says. Victoria, where both road access and London Transport services are badly congested, could only be extended by taking extra land and closing Hugh Street which runs along the south east edge of the station. Docklands is too far from the centre, would need a new river crossing, and would involve considerable property demolition.

The Waterloo scheme, however, has caused almost as much furore as the A20 improvement. Once again there were demands, rejected by the majority of the House of Commons Committee but for strongly sup-ported by Lord Galpern in the Lords' Committee, for the provision to be dropped from the Bill and a public inquiry held. What is at stake here, according to Lambeth Council and local residents, is another potentially endangered species: the innner city community. The Waterloo area, devastated by wartime bombing, partially cleared for the Festival of Britain, has lost 75 per cent of its population since the turn of the century. Over the last 15 years, however, community activists and the Labour Council have been trying to halt the decline and increase the number living and working locally. The effort is symbolised by Coin Street, the site which locals eventually won, against powerful commercial interests, for a community-based housing scheme. All this is threatened, the council says, if the international station is built, for it will bring pressure for office development, new hotels, the conversion of existing houses to small hotels and holiday lets, and the replacement of shops serving local needs by tourist cafes and gift shops. Although present planning policies reject that kind of development, the council says it might not be able to resist new pressures. Hilton Hotels have already made an application in the area. The Government could in any case force a shift policy to favour London-wide rather than local needs.

British Rail disagrees, arguing that international travellers will go to the traditional hotel areas, that extra trading outlets within the station will cater for travellers' demands, and that development policies are within Lambeth's control. It denies also Lambeth's claim that there will be 20 million extra passengers through the station each year — equal to Gatwick Airport — and 4 million extra taxis, coaches and cars. British Rail's forecast is a 13 million increase in passengers by 2003 — leaving the total below the 1978 level of use more evenly dispersed through the day and week. It admits that on the highest forecasts made, the throughput will be approaching the station's capacity by 2003, but says that on the lower forecasts there is capacity for 30 years' growth. It says there will be a 2 million, or 5 per cent, increase in road vehicles and that a new traffic circulation system already introduced as well as improvements planned for Parliament Square will prevent the congestion on local roads and exhaust pollution levels above international standards predicted by Lambeth.

British Rail says that without a single city centre location to help it compete, fewer passengers will transfer from air services and the economic case for rail services would be at risk. Lambeth's case against Waterloo forms part of a larger campaign by environmental groups and councils who want the benefits of the Tunnel more widely spread by, for example, a multiplicity of London terminals and more through-trains from provincial centres. The issue should certainly be seen in the context of increasing traffic congestion in the capital and current uncertainty over future transport policy and facilities in the London region. Labour Lambeth's fears about Waterloo are replicated in Conservative Westminster over plans for a new intercity coach terminal at Paddington. The Paddington plan has broadened support for the idea favoured by the former Greater London Council of using the M25 as a basis of transport interchanges and park-and-ride facilities which would keep commuters' cars and through-passengers out of London. Meanwhile Government-commissioned studies on relieving congestion in four major London traffic corridors are still in progress.

Other Tunnel impacts in London include the use of the West London line, which runs through Earls Court and Olympia, by passenger and freight trains and for the night-time movement of Waterloo-based passenger trains to a new servicing depot on existing railway land near Wormwood Scrubs. British Rail has sought to assuage fears of increased noise and vibration by promising continuous welded tracks, which are quieter and smoother, and electrification to reduce locomotive noise. There is no statutory provision for compensation or insulation against noise arising from increased use of an existing railway, and the House of Commons Committee rejected a plea by local authorities and residents that special provision be made in this instance. The potential problems of noise and vibration are still being studied, however, and some residents may qualify for insulation under plans for a new relief road alongside the railway.

The North Pole site at Wormwood Scrubs straddles three London boroughs: Ealing, Hammersmith and Fulham, and Kensington and Chelsea. Ealing and Hammersmith petitioned for the removal of the scheme from the Bill. Part of the site is zoned for industrial development by Hammersmith which is anxious to use it for creating a larger number of jobs that the depot will provide. Another narrow section, some 6.8 hectares, known as Scrubs Wood has developed into a wildlife sanctuary which the councils and the London Wildlife Trust (LWT) want to

Pets are kept in quarantine kennels near Castle Hill in Folkestone. Many such centres along with a multitude of preventative measures to stop animals ever reaching the Tunnel should allay the fear of rabies.

conserve. The site is a mixture of wood, scrub, grass and wetland on which grasshoppers and lizards abound, with voles and foxes, butterflies, and a wide variety of birds and plant species seldom found in an inner city area. It is regarded as an important ecology-teaching resource for local schools and as a visually important backdrop to the adjoining Wormwood Scrubs public open space. British Rail has promised to preserve what it can of the site as a nature reserve but the London Wildlife Trust says a smaller area will not stand the sort of field studies that the site as a whole can safely carry. The Commons gave a modicum of protection to the wood, writing into the Bill an obligation on British Rail to consult Hammersmith Council and the Wildlife Trust before starting work, when a local schoolboy of 16, Lester Holloway, presented his MP with a file of 144 documented visits he had made to the wood to study its wildlife.

Where will it end?

A senior official from the Department of Transport was walking over the Kentish sites during the Treaty negotiations with a French colleague, explaining that in large areas of the South East conservation interests might resist the development generated by the crossing. "Don't worry about it", came the reply. "Channel it all across to us."

Many in Britain fear that the Nord-Pas de Calais region, with development land to spare, massive public investment in new transport infrastructure, and as a determination to win, will grasp all the economic growth flowing from the Tunnel. Many hope that growth will be generated in Britain but that it will be directed to the depressed areas of Wales, the Midlands, the North, and Scotland. That will depend to some extent on arrangements not yet agreed for customs' clearance on board international trains. It will also depend on the establishment of inland customs' clearance depots for freight in the provincial centres, and on a Government commitment to regional development measures, including public expenditure, far stronger than anything currently on offer. Between the hopes and fears, the reality may be that UK development will concentrate in Kent and the rest of the South East. Market preference will dictate locations where industrial, commmercial and housing development pressures are already strongest: near the M25 ring and other motorways, in the green belt, and around the more prosperous urban areas. Reluctance by the Government and local authorities to let the French take all the advantages could prompt a loosening of existing planning constraints to accommodate these pressures.

The interim report of the Kent Impact Study [20], carried out by the Channel Tunnel Joint Consultative Committee, seeks to assuage fears that the county will be concreted over by new development, arguing that there will be no fundamental change in its economic geography. A threefold increase in cross-Channel traffic since 1970 has not stimulated much growth, it points out. The main focus of the study, however, is not on generating or coping with massive consequential growth but on mitigating the economic disadvantage to East Kent in general and the ferry ports in particular. There is emphasis on tapping the tourist potential, through improved hotel and other facilities, of a European population of 350 million brought within four hours' travel time.

Increased tourism is not without its environmental implications, as the North Downs Society reminded the House of Commons Committee. In this "hidden, mysterious country of woods, deep valleys....tiny remote villages....open sheep walks where you can see for miles" the Society's chairman Mr Michael Nightingale said, old fashioned inns were already attracting coachloads of continental visitors. Meanwhile farmers were busy erecting Community Agricultural Policy intervention warehouses and sending juggernauts down the winding, hedge-girt lanes. The Society's plea, for a special planning policy for the Downs to prevent an invasion by contractors' traffic during the construction of the Tunnel and by tourists and others afterwards, reflects wider apprehension in the county. The Impact Study is not dealing with the environmental aspects of consequential development. That is being left to the local planning authorities, backed by Government assurances that existing countryside protection policies will be maintained.

The study does draw attention to the development potential of Ashford, the only town in Kent signposted for growth in the current country structure plan. It also identifies the likelihood of market pressures on western and central parts of the country now largely protected by the green belt, and the possibility of North Kent becoming more attractive to developers. But its interim analysis is superficial. And while it does concede that the impact on Kent cannot be viewed in isolation, the scope of its work is artificially bounded by the county borders. There is insufficient official examination, in the wider regional or national context, of the potential beneficial and adverse consequences of the Tunnel. A strategy should be devised to maximise economic gains and minimise environmental costs, in addition to the work carried out so far by the Department of Trade and Industry in collaboration with chambers of commerce throughout the UK. Eurotunnel has also employed consultants to send questionnaires on business location and travel requirements to companies all over Britain.

The current development strategy for the South East, drawn up by Serplan, a voluntary consortium of local planning authorities, and endorsed by the Government, is to divert industrial, commercial and housing pressures from the prosperous and now overheated area west of London to the rundown eastern area. A recent study by Serplan [21] has identified development sites on unused and derelict land in the Eastern Thames Corridor, stretching through the East End of London to Southend and Sheerness, and outlined the infrastructure investment and environmental improvements needed to realise their potential. The study was not designed with the Tunnel in mind, however, and makes only two fleeting references to its possible effects. Moreover, Serplan has no special powers to implement the strategy: it can only rely on individual authorities applying restrictive planning policies where development is not wanted and on efforts to persuade the Government to fund infrastructure and other promotional initiatives in designed growth zones.

A study commissioned by the Council for the Protection of Rural England to support evidence to the House of Lords Select Committee points to a series of recent initiatives — the City Big Bang, London Docklands Development Corporation, the Doklands STOLport, the planned new Dartford crossing and the completion of the M25 and other motorways, the North Kent Enterprise Zones and the expansion of Stansted Airport — which are likely to combine with the Tunnel to

Drilling above the tunnel portal-to-be at Castle Hill in March 1987. Dollands Moor where the terminal will be, is on the right, the M20 is in the centre and the light industrial sprawl of Folkestone is to the left of the picture.

generate growth east of London, adding perhaps 80,000 jobs to those already anticipated by the planning authorities. In addition to the east–west growth axis favoured by Serplan, the study identifies a north–south corridor being more attractive to the high growth sector of business and science parks than the derelict sites in the east–west corridor. It fears that pressure will focus around Maidstone, formerly an official growth area for Kent which the County Council is now trying to stabilise by steering high housing demand to Ashford and elsewhere. It believes Kent's faith that Ashford will accommodate the likely demand is misplaced, and suspects that housing pressure will focus on small towns and villages. The Council for the Protection of Rural England and other amenity bodies pressed in the House of Lords for the Bill to require the Environment and Transport Secretaries to make a special report to Parliament, within six months of enactment, on the prospects for growth and the provision of environmental safeguards to contain and counter-balance development pressures. In principle this will occur but there is some leeway on the timescale requested.

Even with the concerted effort urged by Serplan and the Council for the Protection of Rural England to improve access and the environment in the Thames Corridor, it must be questionable whether developers will readily choose the sort of sites available there. London Docklands can exert a strong pull not only because of proximity to the City but also because of the opportunities for an exciting environment presented by the river frontage and striking dockland architecture. Elsewhere perhaps only Chatham, with its enterprise zone based on the old naval dockyard, has that kind of advantage. Abandoned gasworks, derelict industrial sites, claypits and riparian marshes may never be as attractive, no matter how much public money is poured into them. Nor is such a strategy without its environmental costs. The necessary road schemes could draw heavy opposition: witness the public inquiry, lasting more than a year, into the proposed East London River Crossing. And derelict urban sites have their own conservation value, as we have seen with Scrubs Wood.

The Channel Tunnel is a project of immense magnitude and complexity: a package of schemes, many of which in themselves constitute by any ordinary scale a major development with profound and widespread implications. There is a sense of outrage among many of those concerned with the planning and environment of Britain that such a project should have been dragged with such haste through authorisation procedures, with so little time for consideration of the environmental, economic and social implications which ought to set the context for that authorisation. A counter-balanced view is that the project has been discussed for more than 200 years and environmental implications were studied in detail for

the 1974 scheme. Many feel that the Government imposed an unre-
alistically tight timetable on the proceedings and that the House of
Commons Select Committee, in the words of Mr Robin Grove-White,
director of the Council for the Protection of Rural England, acted in a
"cavalier, superficial and unfair" manner, leaving much of the likely
environmental damage unexamined. The Council for the Protection of
Rural England backed its claims — or, critics might say, demonstrated the
effectiveness of its campaign — with a Gallup Poll taken after the
Commons' hearings which showed that 81 per cent of those questioned in
the South East thought there had not been enough public discussion
about the impact on the environment, 62 per cent thought it would
increase the concentration of development in the region, and 66 per cent
thought the lack of a public inquiry set a worrying precedent, but a
surprisingly high number of people wanted the Tunnel despite these
worries. The dissatisfaction is shared by bodies which are essentially
opposed to the scheme, and those, like the Town and Country Planning
Association, who believe it could be a great force for good if its
consequences were properly taken into account and the policy measures
needed to realise its potential were put into action.

The outrage stems initially from the fact that under a Hybrid Bill the
promoters were not obliged, as they would have been at a public inquiry,
to justify their proposals. A public inquiry would not necessarily have
altered the decision to allow the scheme, but it would have permitted a far
more rigorous examination of the need for a fixed link, the type of link,
the route adopted, and the measures required to maximise the benefits it
can bring and minimise the disturbace it will cause. It would have
allowed, albeit at enormous length, a more thorough examination of
every aspect of the scheme. It could, at least, have led to significant
differences in the final product. The Government believes that the time
taken by such an inquiry would have killed the chances of a private
enterprise scheme ever reaching fruition. Given the nature of the scheme,
with its mixture of public and private interests, a Hybrid Bill was
unavoidable. With the exception of the road elements, none of its parts
could properly have been left to customary procedures outside the Bill.
Nor does it seem likely that the planning and environmental aspects
could have been handled concurrently or consecutively by public inquiry
without endless and confusing duplication.

Few if any of those who have been involved as principal objectors in
major public inquiries would disagree that the procedure is unsatisfactory
and needs reconstituting in a more workable and equitable form. Much
public debate and detailed investigation have so far failed to resolve that
problem. Few are likely to say that hearings before Select Committees of

Parliament, in their present form, are an acceptable substitute, particularly for cases of such magnitude. It is to be hoped that the current examination by Joint Committee of Parliament of the Private and Hybrid Bill procedure will help define what is the proper scope of legislation and what should be dealt with by other procedures outside Westminster.

If the Tunnel goes ahead, much work remains to be done on the immediate, the widespread and the long term implications. If the Government does not take up the task, it will lie with other public and voluntary agencies to examine the problems and opportunities, devise solutions, and press for the necessary public action. There remains, meanwhile, the suspicion that the Hybrid Bill procedures will prove irresistably attractive to a Government anxious to hasten the process of economic development by curbing the tongues of both individual objectors and organised campaigners. In both the physical and the procedural spheres, it seems, the fixed link may prove to be the start of a chain reaction.

References

[1] House of Commons Select Committee on the Channel Tunnel Bill, Special Report (CSC.R) para 83.
[2] Commons Select Committee, Minutes of Evidence (CSC.E),p. 803.
[3] CSC.E, p. 421.
[4] CSC.E, p. 444.
[5] CSC.E, p. 251.
[6] Lords Select Committee, Minutes of Evidence (LSC.E) 12 March 1987.
[7] LSC.E 5 March 1987.
[8] The Council for the Protection of Rural England led a group of amenity and community bodies which appeared jointly before the Select Committees. It included the CPRE Kent Branch, Kent Association of Parish Councils, Kent Federation of Amenity Societies, Kent Trust for Nature Conservation, Weald of Kent Preservation Society, and the Ramblers' Association.
[9] *The Channel Tunnel Project: Environmental Effects in the UK.*
[10] CSC.R Appendix 25 and 26.
[11] CSC.R App. 27.
[12] CSC.R App. 35.
[13] CSC.R App. 43, Channel Tunnel Bill clause 8 and Schedule 2A.
[14] CSC.R App. 5.
[15] SCS.R App. 34.
[16] *Motorway Madness*, Friends of the Earth.
[17] *Fixed Channel Link: Impact on Kent's Transport Systems*, Kent CC.

[18] CSC.E P. 801.
[19] CSC.E P. 802.
[20] *Kent Impact Study: a Preliminary Assessment*, Dpt. of Transport.
[21] *Development Potential in the East Thames Corridor*, London and SE Regional Planning Conference.
[22] *The Channel Tunnel and the London Region*, CPRE.

6

Economics and Employment in the Nord-Pas de Calais, Kent and Beyond

Kathy Watson

Tunnel construction, if all goes well, is likely to yield a bonanza for the construction companies involved, but Kent, Nord-Pas de Calais and other areas linked to the rail network could also gain significantly from the scheme.

The county of Kent, in the extreme south east of England, borders Greater London, Surrey and East Sussex on its western side and the Channel to the east. It contains the cluster of ferry ports of Folkestone, Dover and Ramsgate, as well as the Medway area ports, the University of Kent at Canterbury and generally light or service industries.

Current development plans across the 14 districts in Kent are intended to focus development at Ashford in the north, the county's designated growth point. Economic growth is also being encouraged in the urban areas of north Kent and in the east Kent coastal towns, but it is being restrained in the Green Belt areas of West Kent and Canterbury, and stringent countryside conservation policies apply to the open coast and much of rural Kent.

Both Kent and the Nord-Pas de Calais region in north west France which lies closest to the Tunnel will undergo significant changes as a result of its construction. Some of them will be transitory, generated by the influx of men and machines during the seven-year building boom.

Others will be more enduring, as additional roads and rail links are tied into the scheme and the areas become even more accessible. The third set of changes, permanently altering the industrial and economic profiles of the regions, are in the course of negotiation, as each side of the Channel competes to attract much needed industrial enterprise to regenerate its economy.

Ostensibly, Kent would not appear to fit this description. It is after all located in the prosperous south east corner of England, an area generally presumed to have avoided the industrial blight currently afflicting the Midlands, the North, Wales and Scotland.

These assumptions are in fact misleading. Descriptive tags like 'the garden of England' belie the real economic decline that the area has suffered. Between 1978 and 1983 Kent's economy experienced a loss of 19,000 jobs*1 and the closure of Chatham Docks and cutbacks in port-related activities came as a blow in an area whose economy is heavily dependent on maritime activities , and which now clings to the ferries as the prime employer in parts of Kent.

Economic activity in the area is diverse. According to statistical analysis the area can be divided into three main sectors. They are: north Kent with 36 per cent of employment in manufacturing (an unusually high proportion for the south east); mid and south west Kent with a concentration of employment in the services sector (68.8 per cent — a pattern more typical for the south east); and east Kent with a diverse pattern of employment. In this area Ashford has nearly 30 per cent of its employment in manufacturing, Dover approaching 40 per cent in transport (ferries) and Deal has over 25 per cent in mining (all 1981 figures).

Unemployment rates in Kent mirror the wide diversity in economic activity. In 1986, the levels were: mid and south west Kent — 7.9 per cent; north Kent — 12.7 per cent; and East Kent (including Swale Borough) — 15 per cent. This was at a time when the rate for Great Britain was running at 13.6 per cent. But even within these figures, east Kent demonstrates wider diversities — Ashford registered 12 per cent, while Thanet topped the chart with 21.4 per cent unemployment (ibid).

The Nord-Pas de Calais region in north-east France where the Tunnel will emerge lies at the crossroads between London, Brussels and Rotterdam, Paris and Frankfurt. It includes the three ferry ports of Dunkerque, Calais and Boulogne, and the city of Lille. Within the region it has a higher population than Kent — 3.9 million against Kent's 1.45 million, and its unemployment is higher registering an average unem-ployment rate of 13 per cent with a high of 18 per cent compared with Kent's average of 11.8 per cent and high of 20.9 per cent, at December 1985. Overall, the area's unemployment is two percentage points higher than the French average, while Kent's is almost two and a half percentage points lower than the British average. (Figures provided by Kent Economic Development Board.)

In terms of economic activity, the Nord-Pas de Calais region is similar to the north of England. Its economy was based on traditional heavy industry, including coal, steel and textiles, all of which have declined substantially in the last 20 years. But that industrial base is likely to gain it a greater share of Tunnel orders for the French side than Kent will win for the UK side of the operation.

Because this region of France is underdeveloped, and has the potential to gain significantly from Tunnel-related economic activity, the French Government is eager to promote the scheme as an additional spur to its already considerable package of economic incentives. These funding packages are in marked contrast to what is on offer in Kent. There are three main elements to the nationally-financed regional aid package. They comprise regional policy grants, project-related job-creation grants and local business tax concessions. A further package of just under £700 million of aid was committed to the Nord-Pas de Calais region in October 1985.

In addition, the area attracts European Community funding from European Regional Development Funds and European Investment Bank loans (in 1984 totalling just under £29 million). A further range of EEC measures have generated at least £104 million between 1975 and 1984.

In contrast Kent has not benefited from any EEC assistance other than £32,000 from the European Social Fund in the year 1984/85.

The key financial aid within the French regional incentive package is PAT — Prime d'Amenagement du Territiore (ibid). Introduced in May 1982 when the French completely revised their regional incentive system, it comprises a project-related capital grant for manufacturing firms as well as research and service sector companies.

For manufacturing industry, PAT aid is available for setting up and extension costs, for internal conversion entailing a significant change of product requiring major investment, and for relocation costs. They only qualify if the move is from outside to within a designated PAT, but relocation within PAT zones is sometimes allowed. Grant aid can also be provided where a company is taken over in certain situations. All projects qualifying for PAT aid must be assessed as being financially viable.

Most of the Nord-Pas de Calais region is classified as a standard rate zone but individual pockets are eligible for maximum rate support. In the latter, grants are offered up to £4350 for each job created up to a ceiling of 25 per cent of the eligible project investment. In standard rate zones the upper figure is £3050 for each job created, subject to a ceiling of 17 per cent of the eligible project investment.

Transmanche Link's perspective of itself. The group of five British and five French contractors sent out this Christmas card at the end of 1986... the digging has started on each side of the Channel; shovel or spade, wine or beer, ... what will they decide to drink when they meet, mid-Channel?

In assessing the value of the project investment, infrastructure, building and land costs are allowable, but increases in working capital requirements are not.

Those areas of France with especially serious unemployment problems arising from industrial restructuring are eligible for further loans, additional to PAT. For service and research projects the region is eligible throughout for standard rate PAT grants with ten centres (including Lille) attracting higher rate grant aid. In addition, there is linkage to a ceiling percentage of the total eligible project investment for service and research projects, in contrast to comparable standards for manufacturing industry. Service and research projects are ineligible for complementary subordinated loans.

For taxation purposes, aid received under the PAT scheme is regarded as income, but is brought into taxable profits only, over the life of the asset being aided in accordance with the depreciation schedule of the asset. For non-depreciable assets such as land, the PAT grant is divided into ten equal instalments for tax purposes with tax paid over ten years. The standard rate of corporation tax is higher in France than in the UK — 50 per cent compared with the UK's 40 per cent.

Grant expenditure in 1983 and 1984 under the PAT scheme totalled £8.35 million. It comprised 125 individual grants and created just under 4,000 new jobs — an average subsidy for each job created of around £2135. For 1985 grant aid for the Nord-Pas de Calais region was increased to £6 million.

The second major financial incentive, introduced in September 1982, is the regional employment grant — Prime Regionale à l'Emploi (PRE). It is a project-related, job creation grant, administered at regional level and the authorities in the regions are free to specify eligible activities. Most manufacturing and service industries qualify, but agriculture, extractive industries, construction and the utilities do not.

In the Nord-Pas de Calais region, the standard award ceiling is £1740 for each job created — a low figure compared to other parts of France where rural, upland and mountainous areas may attract twice that sum. But in conurbations with populations of more than 100,000, the rate of award is reduced to £870 per job, trapping four of the Nord-Pas de Calais regions — Calais; Lille, Roubaix and Tourcoing; Lens and Douai; and Valenciennes.

PRE grants are financed by the local region and grant aid is only available to cover the first 30 jobs created or, in certain circumstances, maintained. Awards made under the scheme cannot be combined with PAT grants but can do so with other regional aid packages, such as the regional setting-up grant. Because the PRE scheme is administered locally, there is little information of the breakdown of the grant by region. The third French-funded aid programme is the Local Business Tax Concession — Exoneration de la Taxe Professionelle (LBTC). This has been in operation since 1976 with major amendments grafted on in 1983 and 1984. To understand it, it is necessary to chart the basic structure of local business tax (Taxe Professionelle). This tax is paid in quarterly instalments by all businesses, with a few exemptions, to three local centres — the "Commune", "Departement" and "Region". Each of them is responsible for its own portion of the tax, whose rate varies widely throughout France.

It comprises three main elements. First, there is taxation based on 18 per cent of a firm's salary bill in the locality, and secondly taxation based on the company's stock of fixed assets. The standard taxation basis takes the full rental value of fixed assets into account, but where additions to plant, machinery and equipment have been made after the beginning of 1983, only 50 per cent of the rental value is assessed.

Changes in 1983 and 1984 added a further two measures. From July 1983, firms which were set up before the end of 1986, or which took over a company in difficulty during this period, could gain total exemption from local business tax for up to two years from the date that the new firm was established. Further, firms who fulfilled these criteria could also gain a reduction in property transfer tax if they applied. This would be generated via a reduction of between 2 per cent and 13.8 per cent in Land Registration Tax (normally levied at 16.6 per cent on the transfer value of the land).

Sectors eligible to apply include manufacturing industry (but not nationalised or semi-public firms), those involved in the manufacture of pre-fabricated units for the construction industry, and service industries not dependent on local markets and which have a choice of location. Projects which qualify include industrial setting-up of extension schemes, together with those involving major internal reorganisations and certain takeover situations, as well as industrial relocations, particularly those involving decentralisation from Greater Paris or Lyon, as well as from other parts of France. In order to qualify, the schemes must be financially viable and not result in job losses elsewhere. They must be completed in three years and meet minimum job and investment targets, which excludes smaller schemes. The project investment floor for the sevices sector is pitched lower than for investment in the industrial sector, £8700 as against £26,000.

The maximum award under LBTC is 100 per cent exemption from local business tax liability for five years. One drawback for qualifying firms is that if granted LBTC, their corporation tax liability will be increased, since local business taxes are normally deductible from corporation tax.

This wide ranging financial package on offer to firms in the Nord-Pas de Calais region by the French Government, with considerable local discretion and unequalled in Kent, is supplemented by two further types of grant aid, one specific to the region, the other through the provision of European Community grants.

In October 1985, the French Government announced a set of measures targeted specifically at the north west region of France — the "Plan Fabius Pour le Nord-Pas de Calais" following negotiations between the regional council and Central Government. They cover modernisation of industry, new training measures and an infrastructure scheme.

The major components of the package were:

- Support for the Unimetal Rolling Mill at Trith Saint-Leger until 770 alternative jobs were created. As a result, a nationalised industry Thompson CSF is investing around £40 million in an electronic component plant in the area, which will create 600 permanent jobs.
- EDF (Electricite de France) and Air Liquide are to invest around £7 million to establish an electrolysis plant to produce hydrogen for the Ariane rocket programme.

- a new scheme for the partial reimbursment of social security charges relating to newly-created industrial jobs on a reducing scale, fixed at 30 per cent of gross salary in the first year, dropping by 10 per cent in each of the subsequent two years. It will apply only to those areas most affected by industrial restructuring — the Lille conurbation, particularly Roubaix and Tourcoing, the Valenciennes area, the Sambre Valley, the coastal strip and traditional mining areas. Its target is to create a further 10,000 jobs.

- a funding package of £13 million for the expansion of SODINOR, a company engaged in industrial conversion or restructuring. Since inception in 1982, it has enabled 3800 jobs to be created and additional funding is likely to add a further 5000 jobs.

- an increase in funding under the PAT (regional policy grant) scheme (see earlier) to £6 million

- an increase in the Regional Productivity Fund to £13 million for 1986 plus further funding from the EEC which is likely to trigger industrial investment of at least £87 million. This fund is intended to encourage traditional industries to modernise and introduce new technology.

New training measures proposed under the "Plan Fabius" include a new University of Technology in Lille, two University Technological Institutes, one at Valenciennes specialising in electronics, the second at Lens, the National Institute of Applied Sciences. In addition 20 new university degree courses were proposed catering for 500 students, covering maintenance of automated systems, sales and commercial representation and the maintenance of electronic audio-visual systems. A further ten advanced technical degree courses were also planned.

On the infrastructure side, the "Plan Fabius" gave approval to the construction of an express highway along the coast from the Belgian frontier through to Rouen and Le Havre. Other schemes to be accelerated were the spur road to Calais-Marck, and the section of the A26 motorway from Calais to Nordausque (near Saint Omer). On the rail side the plan confirmed the go-ahead for a high-speed rail link from Cologne to Paris via Brussels, with a spur to the Channel Tunnel. These measures are similar to the road and rail schemes planned for Kent as a result of the Tunnel's construction.

A further package of measures announced at the same time in France included the relocation of the SNCF structures and materials testing centre to Valenciennes, improvements to the quayside handling facilities for steel producers and the introduction to warehousing with freeport status at Dunkerque, and the construction of new port facilities for Usinor.

Children play in a sea of balls on Viking 2, a Sally Line vessel from Ramsgate. The shipping company thinks that supermarkets, play areas, pleasure trips and similar devices will keep the other shipping companies and the Channel Tunnel on its toes.

In contrast to the national and regional packages on offer to north-west France, Kent can boast only its five enterprise zones offering capital allowances of up to 100 per cent for firms setting up in the area. Ashford offers loans for new industrial development of council-owned land and also housing for key workers. In addition a small amount of money is available to attract new firms from Kent Economic Development Board. While these measures go some way to attracting new industry, they are minor when compared to the plethora of grant-aid offered by French regional and national Government.

The House of Commons Select Committee on the Channel Tunnel Bill was sufficiently concerned at the inbalance by establishing enterprise zones in East Kent which could produce a distorting effect on less prosperous parts of the country, it added:

"The 'pull' on the south-east within this country is mirrored by the concern of many in the south-east about the 'pull' of north-east France, which as a relatively deprived region of that country enjoys substantial economic support from its Government. We agree that if employers in the south-east were tempted by the Tunnel to move to north-east France in order to receive French Government grants while still having easy access to British markets, it could have a wholly undesirable effect. We look to the Government to ensure that support for industry in north-east France remains within the bounds of existing European Community rules."

The Government's response was lukewarm. In its published reply to the Select Committee recommendations it said, "Although enforcement of Community rules rests in the first place with the Commission, the UK government will certainly consider intervention to support or press for action by the Commission if complaints are brought and there is evidence that EC rules are being breached."

Disparities between the relative merits of both Nord-Pas de Calais and Kent were assessed by chartered surveyors, Debenham Tewson and Chinnocks, in 1986. The authors concluded that the French region would be better placed to attract investment over the long term because of its central position, the generous incentives on offer and the lack of planning constraint.

There is some evidence that north-west France's aid package may be reined in following the recent change of government and the possibility of infringements of EC rules.

Sous La Manche

Channel Tunnellers are intent after more than 200 years' delay to complete a link beneath the Channel or La Manche, not because it represents a technical challenge, which on the whole it does not, nor for innate philanthropy (they are businessmen first and foremost), and not because they see the experience as a valuable addition to their curriculum vitae (although they do), but because, in the words of one participant, "it's like printing money".

Few commentators of the scheme realise the extent to which the Channel Tunnel represents a major opportunity for the builders involved to make substantial financial gain, as long as they keep their nerve and retain the confidence of the politicians.

The ten contractors who will build the scheme, Transmanche Link (TML), are intent on doing that, despite the political storms that have broken over their heads.

TML is an amalgam between Translink and Transmanche Construction. The first is a joint venture between British contractors Balfour Beatty Construction, Costain Civil Engineering, Tarmac Construction, Taylor Woodrow Construction and Wimpey Major Projects. The second is made up of French contractors Bouygues, Dumez, Société Auxiliaire d'Enterprises, Société Générale d'Enterprises and Spie Batignolles.

For the UK Government the scheme has always been a winner. Faced with the seeming insoluble problems of rising unemployment and while holding to the philosophy of maintaining a tight hold on the purse strings, they have enthusiastically promoted the project which will provide the nation with a major piece of infrastructure on the cheap. Their argument runs that it will generate jobs nationwide during construction, reinvigorate the skewered economy of Kent, and, in the long term, provide UK manufacturers with a cheaper route into the markets of Europe while encouraging overseas' investment in the UK and pulling in tourists — all at little cost to the taxpayer. An added bonus is that eventually ownership of the rail tunnel, together with the road tunnel, if it is ever built, will revert to the public domain.

For the contractors the potential benefits far outweigh the risks. The scheme provides them with a highly lucrative construction contract as long as they can keep to the timetable. Once operational, the Tunnel has almost 50 years in which to claw back the investment of the concessionaires, which include the British and French builders as founder shareholders.

The Tunnel itself is designed to carry not only passengers, vehicles and freight, but also fibre optic cables under a deal being worked out with British Telecom, and possibly an electric cable to bring French electricity, generated by their nuclear power industry, through to the UK grid, and should the economic picture change, to send British electricity to France.

All these secondary deals, with la SNCF and British Rail, British Telecom, the Electricite de France and Central Electricity Generating Board, and freight exporters will generate revenue in addition to what the Tunnel earns from its main traffic — passengers.

There are also tertiary spin-offs. The contractors have linked to councils and authorities local to the Tunnel portal to form Kent Economic Enterprises Ltd — a group formed to identify and take advantage of business opportunities spawned by the Tunnel. Both tourism and leisure have been highlighted so far, and local councils are keen to take up offers from promoters, with input from the contractors. Currently the Carroll Group is proposing a major leisure scheme at Ashford, the £150 million Eurocentre, the area hallmarked for growth in Kent, and Dover Council is keen to boost its heritage facilities. Plans for hotels, leisure complexes and regional conference centres are beginning to rain like confetti, upon local authorities whose initial ambivalence to the scheme is converting to enthusiastic support.

Although it may seem odd to some that building contractors should move into transport operations, it is part of a growing trend of diversification. Following the downturn in construction through the 1970s and 1980s as public expenditure on infrastructure plummeted, contractors faced with returns as low as 2 per cent on construction activities have been actively pursuing the chance to broaden their operating base with notable success.

Construction giant, John Mowlem plc, took the lead by promoting a short take-off and landing airport (STOLport), now renamed London City Airport(LCA), in London's Docklands. It was a shrewd gamble which pre-dated the area's current vigorous growth and the announcement of the massive financial centre to be built in Canary Wharf.

Mowlem carried out the construction of LCA and will operate it when the first commercial passengers start arriving in autumn 1987. The firm is likely to repeat the exercise in the docklands of other major cities, such as Sheffield and Cardiff.

In 1987 Trafalgar House (the earlier promoter of EuroRoute) was rebuffed in its plans to build and operate a second crossing of the River Severn. But it succeeded in winning Government approval for its scheme for the Dartford crossing, a bridge across the Thames which it will build and operate before eventual handover to Government.

Further innovative schemes have been put forward for privately financed energy gathering barrages across the Mersey and Severn. Construction firms are hopeful that, if the Channel Tunnel scheme proves successful, it will demonstrate their capability in many spheres.

Since the late 1960s contractors have chivied successive governments for a chance to build and operate major road schemes, using tolls to recover construction costs. Their schemes were not well received when time and time again the inefficiency of tolled routes was demonstrated. But there can be little doubt that the Channel Tunnel scheme has breached onetime Government opposition to private funding and operation of major pieces of infrastructure. The approval for the Dartford bridge would have been less likely without the contractors' Tunnel scheme first gaining politicians' support.

The expertise that UK contractors stand to gain can be put to good use when they chase contracts overseas. So far they have felt handicapped in the national competition for major contracts by their lack of demonstrable experience in funding, building and operating large developments in the home market. The Channel Tunnel goes some way to redressing that shortfall.

For the construction industry as as whole there are considerable spin-offs in work which the fixed link will generate.

For its part TML, the Tunnel contractor, is committed to building two main tunnels and a service tunnel, with rail tracks, each about 50 kilometres long, comprising four kilometres under the French mainland.

Tourism will be a vital part of the future economy of the Nord–Pas de Calais region. This picturesque church is not an isolated beauty spot but one of the many attractive features in the countryside of woodlands and hedgerows.

In addition, they will construct the Cheriton terminal at Folkestone, an inland clearance depot at Ashford and related work including the extension of the working platform below Shakespeare Cliff, west of Dover and a similar set of facilities in France.

The terminal facilities will include toll booths, immigration and customs control, petrol filling stations, shops (including duty-free if permitted) and restaurants. In all, the UK buildings will cost £1204 million to build, just over £300 million more than their French counterparts (at September 1985 prices).

On the UK side, land required for these works will be acquired by the Secretary of State, at the concessionaire's expense, and leased back to them over 55 years before handover to Government. Land required only during construction will be returned after ten years to the original owner or, if they did not wish to repurchase or the land had been acquired by agreement, transferred permanently to the Channel Tunnel Group (the UK arm of Eurotunnel, made up of building contractors and banks).

As part of the Tunnel scheme, but separate to the contractors' contract, British Rail plans to build an international terminal at Waterloo by remodelling platforms 16–21 to provide five 400 metre international platforms. A further four will be built to cater for south-east traffic so that existing services are retained. Other British Rail works comprise a new passenger station at Ashford, a flyover on the Stewarts Lane Curve, a servicing/maintenance depot near Wormwood Scrubs, a freight inspection facility at Dollands Moor in Kent and additional freight facilities in Willesden. These, together with smaller items, are costed at £180 million and, together with the £200 million of rolling stock needed, will bring British Rail's bill to £380 million (all at 1985 prices).

French railways (SNCF) plans international passenger terminals at Frethun and Paris Gare du Nord, with Belgian railways improving Brussels Midi, with a freight facility near Calais.

The road schemes necessary to cater for tunnel traffic are the provinces of Kent County Council, and the Department of Transport, and to a lesser extent Sussex County Council.

Kent expects to spend around £50 million on a new road construction including the Ashford Southern Orbital Route (a dual carriageway from the M20 to A2070), the A2070 diversion round Ashford. Three other schemes make up the total: A256 (Sandwich to A2 at Dover); A260 (Barham to A2 Folkestone); and A2070 (Brenzett linking A259 with Ashford).

For the Department of Transport there are four major schemes under consideration costing around £115 million. They include the A20 Folkestone to Court Wood and the continuation from Court Wood to Dover, the M20 "missing link", Maidstone to Ashford, and improvements to the M20 Maidstone Bypass.

The Channel Tunnel Bill provided authority for many of these schemes to go ahead with deemed planning consent although decisions on building design and layout would be reserved for local planning authorities. Further construction work is likely to be generated by new industry moving to Kent, in addition to the leisure and tourist schemes already mentioned. Local authorities have visited the Pacific area in order to trumpet the attractions of Kent as a gateway to Europe for manufacturing industry. Companies from the United States and Japan have responded but because of the long term nature of the commitment, their plans are not yet revealed.

The final major gain for TML derives from the expertise it will have accrued in putting together a major infrastructure package of this magnitude. If it succeeds, it will have demonstrated considerable skill in putting everything from design, construction, through training, funding, commission and operating, as well as dexterity in enabling the locale to take advantage of resulting economic benefits. Such a package would be a strong bargaining weapon in the competition to build a permanent link across the Strait of Gibraltar, the Denmark–Sweden link crossing the

Oresund, and a tunnel to link Italy and Sicily, all of which are currently under discussion.

It is no wonder that the Japanese have been eager to invest in such an opportunity. There are even rumours that acquisitive Japanese contractors now coming into the UK in strength are casting hungry looks in the direction of the Tunnel partners. It is not inconceivable that a Japanese company such as civil engineering contractor Kumagai Gumi, could purchase the likes of Wimpey or Costain, in order to gain a share of the work.

What makes the prospect even more appetising is the fact that British contractors have proved in recent years that they can deliver a scheme on time and within budget. Performance of the French companies has not been in doubt. A study by the National Economic Development Office into the efficiency of the engineering construction section since 1981 has shown a significant improvement in its ability to build within schedule and budget. Of 25 major schemes built within the time period, only one turned in late, five months behind schedule. The rest were completed on time or within a time scale to the client's satisfaction.

It is not often understood outside the construction industry that, while contractors regularly spar against each other in the contest to win contracts, they are accustomed to working together on major schemes in consortia, each handling a discrete section of the work. But the Tunnel scheme is one of the few in which all five UK contractors will work together as a team, pooling their workforces. To date only one major work package is likely to generate a different work method, railway design, in which Balour Beatty, because of its acknowledged expertise, will lead the team. The construction contract is tight, with TML acting as management contractor for the design, construction, testing and commissioning of the Tunnel. It is divided into three parts according to the amount of work and risk involved: the tunnels to be built on a target cost basis; the infrastructure for the terminals and fixed rail elements, on a lump sum basis; and the shuttles and locos on a competitive basis. For details, read Chapter 1. Although construction firms commonly boost their price by submitting claims for additional cash during and after construction, this is unlikely to occur in the case of the Tunnel for two reasons. In this case the contractor is also the designer; therefore the usual procedure whereby an independent designer alters specifications and triggers a claim from the contractor wiil not happen. However, if Eurotunnel elects to make fundamental changes to the scheme — perhaps to accommodate additional services — this will give rise to contractor claims. The only other possible reason for changing the target which might generate a claim is if the contractor hits exceptionally bad ground conditions which an experienced contractor could not have foreseen. This is particularly unlikely on the UK side.

In order to oversee progress, and standing independently of both Eurotunnel and TML, a joint venture of consultants has been appointed as Maitre d'Oeuvre. The firms involved are W. S. Atkins and Partners and the Societe d'Etudes Techniques et Economiques (SETEC), supported by Sir William Halcrow and Partners and Tractionnel Electrobel Engineering. Their job will be to advise Eurotunnel on all aspects of the contract, monitor construction to ensure it meets the standard required, act as arbitrator on costs and provide information to the Intergovernmental Commission and lending banks.

Construction orders

While architects enthusiastically court publicity for their latest schemes, the construction industry is notoriously secretive about its operations, at least until a scheme is sufficiently far advanced with all problems ironed out, at which stage they become more approachable and their building sites more accessible. Their defence is usually that the client prefers it that way. Not so in the case of the Channel Tunnel, where the client, Eurotunnel, has swamped the media with an avalanche of briefing papers, reports and digests. It has chosen this stratagem because of the political nature of the scheme and the need to win public as well as political support in the face of strong opposition from Flexilink (the ferry and harbour operators group) and other groups committed to killing-off the scheme. These papers have included unusually detailed lists of the main construction and civil engineering equipment and materials required); lists of the areas of the UK which are likely to win those orders; and preliminary lists of potential UK suppliers by name and geographic location.

Although at the time of writing it was still early in the contract-awarding process, Eurotunnel's plans to spread work and therefore jobs, throughout the UK have come unstuck on at least one occasion. With the requirement to advertise at least one third of the contracts within the European Community those plans are unlikely to go awry again. At the outset of the scheme as a spur to manufacturers, Eurotunnel compiled the names of over 260 firms from a desk study of those with a track record in the industry. In the west of the UK, 50 firms were nominated, ranging from providers of structural steel work to suspension systems, from tunnelling equipment to ticket-issuing machines. In Scotland, 38 companies were identified as potential suppliers of cranes through to concrete, manhole covers to locomotives. In Wales the figure was 37, in

Valuable contracts for drilling and site investigation crews were awarded well before the main tunnelling could begin. Here a Terresearch platform is seen carrying out sheet piling tests prior to seawall construction at the base of Shakespeare Cliff.

the north west — 52, in the north east it rose to 70 firms while for the Midlands the list of potential suppliers reached 120. At the time of compilation TML stressed that the list was not exhaustive and invited other interested firms to contact them. Overall response has been good.

The requirement to advertise 30 per cent of the contracts in the Official Journal of the European Community presented Eurotunnel with a major problem of how to make sure that UK firms were aware of forthcoming invitations to tender and how to accelerate their responses. After all, UK manufacturers and suppliers, unlike their European counterparts, were unaccustomed to regularly trawling this potentially lucrative source of work. Shudders must have run down the collective Eurotunnel spine at the spectre of overseas' companies winning large parcels of Channel Tunnel work at a time when, with the enabling Bill under consideration in Parliament, the promoters were capitalising politically on the potential job gains for the UK. This danger was averted when the Department of Trade and Industry stepped in with the offer to set up a co-ordinating section in order to disseminate information and supply copies of the latest EC journals to interested firms through its regional offices.

A second major problem hit the contractor over the manufacture of precast concrete linings for the Tunnel. Over 600,000 are needed overall, rising at peak to over 1000 a day. The £100 million order was originally targeted for the Midlands, but TML unexpectedly decided to manufacture the linings itself by setting up a factory on the Isle of Grain in Kent, rather than awarding the contract to either of the two regional consortia who were interested. TML's justification for holding on to this work was that it needed to guarantee an unbroken supply of linings during construction, and any delays in delivery due to British Rail strikes or manufacturers' holdups would have halted all progress on the scheme. One company hopeful of work, Charcon Tunnels, was bought by 20 per cent TML shareholder Costain in April 1987 and so, even if indirectly, it is still likely to obtain some work.

Of the preliminary orders placed for the UK side of the Tunnel, the vast majority have been won by UK firms. The majors include the first order for two tunnel boring machines, valued at £6.4 million to James Howden of Scotland, which is well placed to pick up the rest of that £24 million contract for the remaining four machines for the UK side. Howden had also picked up a £3.7 million order for the construction rolling stock in a joint venture with the West German designer and manufacturer, Muhlhauser. In Leeds, Hunslet Engineering built 12 construction locomotives in a £4.8 million contract. Blue Circle won the first of the cement orders with a £13 million contract to supply the Isle of Grain, close to its Kent base. Foster Yeoman is providing £10 million of aggregates, most from its Glensanda quarry in Scotland.

The first major encroachment by overseas' manufacturers has come from the Italians. Their firm Sacma linked with UK-based Modform systems to win the £6.6 million order for tunnel segment moulds and production line equipment. In 1987 a further batch of contracts exceeding £50 million in value will be placed. Out of the overall total of £700 million orders of the UK end of the Tunnel, £100 million is expected to be won by Kent firms.

Although Tunnel orders have brought unexpected good fortune for hard pressed manufacturers in the UK, they have been quick to realise the dangers inherent in wholeheartedly supporting the scheme. Eurotunnel and TML insist on high quality and fast production and offer bonus schemes for early completion. But because of the dearth of orders in the past, those firms who have managed to survive the recession are unwilling to switch all their production over to Channel Tunnel work. To do so would endanger their goodwill with long-standing clients whose custom will be needed after the Tunnel is complete. It might also provide further footholds in domestic markets to overseas' competitors. Logic dictates that UK manufacturers spread the load between them by linking in joint venture, but because of their declining numbers, many have found it necessary to link up with overseas' firms in order to turn in strong bids, evidenced by the Modform/Sacma and Muhlhauser/Howden associations.

Jobs

The great jobs' debate lies at the heart of much of the Channel Tunnel controversy. Eurotunnel claims that its scheme will generate jobs during construction in Kent and throughout the UK. Once up (or down) and running, the Tunnel will provide permanent work through its own operational needs and the additional customs and excise and British Rail staff to service the system. New jobs will also be created from tunnel spin-offs, the new tourist and leisure facilities to be provided in Kent, port development on Clydeside, hotel staff at Waterloo in London. Opponents of the scheme claim otherwise. They argue that the ferries will be badly affected by competition from the Tunnel and that thousands of jobs will be lost.

Both sides sprinkle their arguments with "ifs" and "possiblys" and, in truth, despite some very sophisticated models being constructed to prove their arguments, it is very difficult to anticipate the full effects of any major infrastructure scheme. No-one knows that better than the queues of frustrated motorists who daily travel on the M25, London's orbital motorway. That scheme was designed to relieve congestion around London but since completion has been on permanent overload to such an extent that, even before the whole road opened, plans were already in the pipeline for its enlargement to cope with the huge volume of traffic it was attracting.

From first gaining the concession to build and operate the Channel Tunnel in early 1986, Eurotunnel let it be known that the construction workers would be recruited from Kent and the south east. Initially this seemed logical, since to recruit men from elsewhere would require time and money for site accommodation. But, over time, because of the political controversy over jobs, it became clear that more political kudos could be gained if some construction workers were recruited away from the "prosperous south east". Further, in the run up to a general election the Government could not be seen to actively support a major job-spinning scheme whose benefits embraced the south but not the north, despite there being several thousand unemployed miners in Kent.

The great north/south divide argument and the debate about the south of England growing in prosperity while the rest of the United Kingdom languished in the doldrums of economic recession featured largely in pre-election political posturing in 1986 and 1987 and spilled over on to the scheme. One of the more stirring speeches in the second reading of the

Channel Tunnel Bill in the House of Lords concentrated on the north/ south divide.

Other forces were also at work to persuade Eurotunnel to rethink recruitment. Most pressing was the realisation that there would be a great deal of competition for construction workers from the other major building schemes setting up at the same time. These include the expansion of Stansted Airport, Trafalgar House's Dartford Bridge, and in London's Docklands, three huge schemes for the Royal Docks and the Canary Wharf financial centre. The French side would also be affected by large construction projects such as the first European Disneyland being built near Paris.

Worries were compounded by the knowledge that Kent's jobless would not necessarily have the skills required for the scheme. Eurotunnel ran a cross-match of those available for work and the skills' requirements and identified nine occupations in short supply. They were:

Civil engineering operatives of various types
Electrical tradesmen
Mechanical tradesmen
Joiners
Scaffolders
Tunnellers — skilled and semi-skilled
Specialist steel workers
Concrete finishers

Although some tradesmen such as miners who had moved to Kent from Wales and the north east in search of work could be retrained as tunnellers, the local workforce was unlikely to be able to fill all the gaps.

For all these reasons, Eurotunnel and TML decided to recruit nation-wide and set up a sophisticated computerised system linked to the Manpower Service Commission's Job Centre network. It has since coped with a huge number of applications from all over the country. The system is so detailed that it can provide the enquirer with the name and address of a tunneller in Tooting, a scaffolder in Solihull or a carpenter in Cleethorpes (as long as they have applied).

Before being employed, candidates will be vetted through the shadowy agencies that serve construction to weed out known trouble-makers. Those who succeed and who need accommodation will be offered space in the construction camp established near the Tunnel workings for around 2100 staff. Others chasing bedspace will be housed in one of three caravan sites, at Westhanger, Hawkinge and within the terminal bound-ary. But any illusions that TML is venturing into the "Hi De Hi" world will be short-lived. The workforce will be on a 24-hour shift system, seven days a week. "But when they are finished they'll be able to retire to

The quiet site at the base of Shakespeare Cliff in 1985. Already in 1987 it is bustling in preparation for work to begin.

the Bahamas" said one envious Eurotunnel penpusher. Dover could become the Aberdeen of the south.

TML intends to train more than 100 apprentices in two full apprentice-ship schemes during the seven year construction phase. They will probably all come from Kent. Although these numbers seem small, they will provide the construction industry with much needed qualified manpower in the face of growing skills' shortages on the burgeoning building sites in the south east. Further training wil be provided through a Local Collaborative Project set up between TML, Kent County Council Education Department and the Manpower Services Commission, which will match Kent's training resources to TML's needs.

The long lead time before construction starts gave TML valuable planning space to prepare workers' facilities, plan sites and draw up working procedures. This may not seem important to most industries, but construction on such a scale has special needs. Where else does the place of work re-locate so regularly? Where else does the size and shape of the workplace change day by day?

This time fencing round the site will be more durable than most, and TML plans to provide good on-site facilities, well above the plank slung across two bags of cement with a bucket round the back set-up that one still sees on small sites. Workers on shifts round the clock will need good accommodation, superior canteens and decent leisure facilities, that do not impinge on local residents as feared, to keep them happy and their productivity up. But keeping them happy has to be one of TML's greatest imperatives. It has a tight construction schedule to meet and any delay in completion, through strikes or any other hold-up, will cost £250,000 per day.

Another major benefit of the long lead time prior to the start of work is that it gives TML time to programme for safety. Few people outside the industry realise how hazardous construction is. The industry usually looks forward with trepidation to the annual report from the UK Health and Safety Inspectorate in which it is regularly lambasted for its poor record.

The problems associated with keeping a site safe are legion. In no other industry is there such a wide mix of places of work: on scaffolding high above ground level, in trenches which can and do collapse, at ground level where workers not only run the risk of being struck by falling objects but face that potentially lethal combination of men and machines in a general free-for-all. All in a landscape that changes daily and in which workers have always to remember that the guardrail that may have been in place yesterday was probably taken away today when its owners finished on site and that the ladder linking the second and third floors was tied in place by a careless sub-contractor.

It would be unfair to imply that all building sites in the UK are badly managed. Most of them, especially the large sites, are well-planned and the industry is fighting to improve safety awareness. But in many cases the message has not yet been received by the workforce. While overseas' construction workers automatically wear head protection on a building site, their British counterparts are less astute. In a predominently young male environment, the macho ethos prevails. But as one member of TML management pointed out: "It's not the kind of job that you win on a Thursday and start work on site the following Monday. With the Tunnel it's different. We have the time to plan and get it right."

TML plans to give all staff half a day's safety training as soon as they arrive, with regular follow -ups. The workforce will be directly employed by TML. More usually, site workers are self-employed and therefore responsible for their own tax and insurance. This practice grew up in the early 1970s as workmen cut loose from contracting firms, preferring to move from site to site in search of jobs. In 1987 almost 50 per cent of UK construction workers were self-employed.

Drawbacks far outweigh the gains. On-site training opportunities have been reduced as the ratios between experienced and inexperienced staff were lost. Firms were able to shirk the expense of employing apprentices.

Site safety is difficult to maintain with an ever-changing workforce, workmanship suffers, and considerable tax revenue is lost.

Weaning the industry off this habit has been a slow process. TML took a long time in its negotiations with the construction unions to secure a mutually acceptable agreement. Both the Transport and General Workers Union (TGWU) and the Union of Construction and Allied Trades and Technicians (UCATT) were keen to seize their opportunity to boost membership by gaining access to the TML workforce. They have both suffered declining membership and reduced power in the industry through the popularity of self-employment.

For TML the wage deal was complicated and was based on the Civil Engineering Working Rule Agreement supplemented by bonus schemes, pitched sufficiently high to draw workers back into direct employment. It needs to be flexible enough to maintain high productivity levels which reward the diligent shift team over the indifferent, while demonstrating fairness in the event of on-site delays.

Since the turbulent times in the 1960s and 1970s when British civil engineering and construction sites regularly hit the headlines as strikes brought work to a halt, the industrial climate has improved. Better site agreements and reduced militancy among the workforce has aided the industry's track record, enabling it to demonstrate a new-found ability to complete work on time and within budget. It is perhaps for this reason that TML in the foyer of its Croydon headquarters proudly sports a calendar which is counting down the days to tunnel completion. In a team as experienced as those heading TML, it augurs well.

Eurotunnel's initial assessment of employment created during construction breaks down into six areas:

(a) 15,300 man years on site preparation, Tunnel construction and work for British Rail.
(b) 1650 man years in East Kent through the supply of cement, concrete products and aggregates.
(c) Away from Kent, up to 60,000 man years in the supply industries arising from the £700 million of orders.
(d) 3628 man years of British Rail direct employment on fixed work in the London area.
(e) 2500 man years of employment in Kent, in retailing and other areas where construction workers spend their wages. This is based on a local employment multiplier of 1.2 — the economist's shorthand for the number of jobs created when the wage level in an area rises — which indicates a further 20 jobs being generated from every 100.

(f) 1,275 man years from consultants and staff of Eurotunnel and TML who are administering development and construction. That represents 19,450 man years of work in Kent and 64,900 in the rest of the country, and a grand total of 84,350 man years.

It is true that quoting figures in man years, rather than by the number of jobs created, is difficult to digest, but it would be misleading to do otherwise. Although construction work on the scheme will last for seven years, jobs will fluctuate. A glimpse of the peak construction year, 1990, illustrates the point.

Just under 4000 workers will be employed in 1990 distributed as follows: 1720 on the tunnel workings at Shakespeare Cliff; 630 on the container depot at Ashford; 230 on the manufacture of tunnel linings on the Isle of Grain; and 1400 in Folkestone. Adding the multiplier effect, the 20 additional jobs created as direct spin-offs for every 100 jobs, the total rises to just under 5000.

But these figures belie fluctuations which will occur during construction. Starting in 1986 with employment on construction at 77, this will rise through 1987 to 1120; 2220 in 1988; 3460 in 1989, peaking at 3980 in 1990 and thereafter falling back — 3730 in 1991; 2310 in 1992 and 770 in 1993. (Source: Eurotunnel).

These job totals do not include jobs created through secondary schemes which the Tunnel will trigger, such as retailing, distribution, and hotel, office and house construction and leisure, during both construction and operation. It has been estimated that during construction, 19,000 jobs will be created in France, of which 11,000 will go to the Nord-Pas de Calais. Eurotunnel has used a multiplier of 1.2 to gauge the additional jobs which will be created while construction is underway in south-east England, but for the rest of the UK the figure is higher: 2, or one additional job for every one that the scheme creates. Following completion and with operation of the fixed link getting under way, Eurotunnel expects a further 30 jobs to be created for every 100 directly employed — a multiplier of 1.3. As before, this does not take account of secondary spin-off schemes.

Initial estimates from TML indicate that the average length of stay for Tunnel workers will be around three years. That labour turnover combined with continual training requirements will keep the Ashford training centre in business over most of the lifetime of the scheme. TML will focus induction training on safety and quality control. New recruits to construction will be given the chance to learn new skills. Other staff will be offered upgrading or refresher courses.

Once the building phase is over, Kent will not be plagued by thousands of construction workers roaming the county without jobs. The average brick layer is not like his counterpart in the steelmaking industry in, say, Corby, who is left unemployable when the works close, to face a bleak

A mass of steel reinforcement and wooden formwork on the French terminal site. Major construction work started there at least six months ahead of their British counterparts.

future. Construction is an itinerant industry and the men go where the work is. In addition, the lay-offs will be gradual and most of the men can expect to be absorbed into the UK contractors who make up TML — Balfour Beatty, Costain, Tarmac, Taylor Woodrow and Wimpey.

While construction of the fixed link progresses, the lucrative tourist market, likely customers of the scheme, will also be expanding. The English Tourist Board expects tourism to grow by 21 per cent in the 1985–91 period especially through the short-break holiday market.

Overseas' tourism to the UK will also be increasing, from 1.45 million visitors to the UK in 1985 to around 20 million in 1992. Not all will use the

Tunnel, but Eurotunnel is hoping to attract the major share.

The operational phase

Once operational, the Tunnel will provide permanent work for both Eurotunnel and British Rail staff as well as the Immigration, Customs and Excise officials needed. Additional jobs will come from the expansion of Kent's economy to cope with the influx of tourists, and both Kent and the whole of the UK will benefit from the economic spin-offs of having a fixed link into the heart of Europe. But, true to the old adage, "you can't make an omelette without breaking eggs", the Tunnel will damage the older established businesses in the ports and among ferry operators in the competition to win passengers and freight traffic.

Eurotunnel expects most of its revenue to come from transporting passengers. In 1993, the first year of business, around 63 per cent of the business will come from passenger traffic, and around 28 per cent from freight (the rest is the 9 per cent ancillary revenue from the possible cable carrying facilities and possibly duty free), representing a total revenue of £389.2 million (at 1985 prices). Ten years later, the picture is expected to change slightly with total revenue rising to $480.6 million, but the proportions changing as the passenger shares moves downwards to 59 per cent and the freight proportion rises to 32 per cent. The Tunnel will generate additional "curiosity traffic". Around 9 per cent of the passenger business will come as a direct result of the Tunnel, made up of people who would not otherwise use cross-Channel facilities.

Eurotunnel's passengers will be drawn from those taking independent journeys to mainland Europe by sea or air, those travelling independently by car and organised leisure trips on coaches by sea. It is reckoned that 60 million people will be within four hours travelling time of the Tunnel, and 100 million within six hours travelling time.

Freight business through the Tunnel will come from container and other bulk movement between the UK and Europe together with traffic between Eire and the Continent which passes through Britain. But British manufacturers have been slow to grasp the advantage that a fixed link will provide in terms of easier and faster transport into the heart of Europe. To overcome this, Eurotunnel organised a roadshow in 1987 to sell the idea to them. Most of those who attended said that although the scheme was seven years away, they would take advantage of enhanced transport facilities and many would start to look afresh at Europe as an export market.

The benefits for freight exporters are considerable. British exporters will gain a fast, reliable service via British Rail which will cut out the double handling element as freight switches between rail/road and ship and reduce the dangers of damage in transit. The service will be more flexible, offering the opportunity to transport more frequently with better customer control over goods. Dangerous freight, such as bulk fuel, will not be taken through the Tunnel because of the inherent risks. Forecasts of the likely proportion of passengers and freight traffic which is likely to choose the Tunnel instead of ship or aeroplane make bleak reading for Eurotunnel's competitors. For those who want to travel by sea between Britain and France, Belgium or the Netherlands, about 60 per cent of those travelling with a car, about 70 per cent of those in coaches, about 75 per cent of foot passengers on an excursion and almost all non-excursion foot passengers are expected to divert to the Tunnel (Source: Eurotunnel — *Pathfinder prospectus* — 1986). Air passengers heading for France, Belgium/ Luxembourg, Holland, West Germany, Italy, Switzerland, Austria and Spain, Portugal, Yugoslavia, Denmark and Greece will also switch to the Tunnel — around 17 per cent of them on Eurotunnel's reckoning, on the assumption that a high-speed railway is built on the Continent.

Unitised (container) freight traffic is also forecast to make the switch. About one third of the traffic currently being shipped by sea is likely to divert to the fixed link. But Eurotunnel will make fewer inroads into the non-unitised freight market, picking up only around 8 per cent of the market excluding tanker cargo. For the lorry-driven cargoes, time savings may be more apparent than real since drivers generally use ferry crossing time as their necessary rest period, but if a decrease in time for customs' clearance was possible for either mode, effects would be more dramatic.

In all, Eurotunnel expects to transport just under 30 million passengers during its first year of operation — the majority, 15.9 million, on through-trains, the remaining 13.8 million on shuttle trains with their vehicles. But as the British pick up the habit of more travelling with their cars, the picture will change.

By the year 2003, Eurotunnel expects to be carrying a further 20 per cent of passengers, the majority, some 20.3 million, will be making the trip in shuttle trains with their vehicles. Train-carried passengers will grow more slowly, rising by under 2 million. For freight traffic the estimated growth between the years 1993 and 2003 is expected to be around the same level. Total freight in 1993 is put at 13.2 million tonnes (six million tonnes on shuttle trains and 7.2 million on through-trains). Ten years later the former is expected to rise by one million tonnes, while the latter, freight carried on through-trains, will grow by 3.4 million to 10.6 million tonnes. These figures of projected market share and passenger and freight growth have been hotly disputed by Tunnel opponents. Their main contention is that the scheme will not succeed in invigorating sluggish markets and Eurotunnel forecasts of traffic share are wildly optimistic. The outcome they claim will be a price war between ferries and the Tunnel because of over-capacity. (Conservatives against the Tunnel, October 1986.)

For those who do choose the Tunnel, British Rail and SNCF plan to offer foot pasengers up to six through-services daily in each direction from west Scotland, north west England and the Midlands, including possible overnight sleepers. The traveller from Glasgow will gain the change of reaching Brussels in 9 hours and Paris in 9 hours 15 minutes. From Birmingham the trips will take 5 hours, and 5 hours 15 minutes respectively. On the east coast main line, passengers from Edinburgh will reach Brussels in 8 hours 45 minutes and Paris in 9 hours, on any of six trains. Wales and west England will be served by up to four connecting trains, as opposed to through-trains, daily in each direction, bringing Cardiff within 5 hours of Brussels and 5 hours 15 minutes of Paris.

The London-based traveller, because of his close proximity to the Continent, benefits most. An hourly fast service for most parts of the day will get him to Brussels in 2 hours 55 minutes and to Paris in 3 hours 15 minutes. This compares with train and ship time of 6 hours 55 minutes and 5 hours 15 minutes by hovercraft from London to Paris, and 6 hours 50 minutes by train and ship and 4 hours 55 minutes by jetfoil to Brussels. Two sleeper trains will run nightly from London via Brussels for north Germany and Switzerland, together with connectons to overnight trains from Frethun to Southern France, Spain and Italy. Many of these trains out of Waterloo will also stop at Ashford. Railfreight customers will be offered up to 12 wagonload services daily in each direction run during off peak periods and for containerised traffic there will be at least eight container trains daily in each direction, including an overnight service from principal UK cities to centres around 800 kilometres from Calais. This should produce a journey time of 36-60 hours to the major European centres, and if siding to siding or inter-terminal services are used, these figures could fall to 24–36 hours. The exceptions are Milan and Madrid which are expected to take three and four days respectively.

Customs' clearance will take place at the international container depots, the originating premises or British Rail Willesden depot.

While the arrival of through-tunnel freight trains may represent bad news for those currently engaged in freight haulage, it is good news for the small towns and villages plagued by juggernaut lorries. British Rail estimates that 1500 lorries a day could switch to its rail facility, arresting the current decline in rail freight. Traffic projections given above are unverifiable until the Tunnel is operational but, in the interim, are moving upwards all the time, even with data emanating from oppositon groups.

Full-size testing rig constructed by a French-owned UK manufacturer of lining segments, seeking part of the lucrative contract to be awarded by Transmanche Link.

When the Tunnel opens for business in early summer 1993, it will be accompanied by a dramatic reduction in current fares, as ferry operators and airlines scramble to retain market share. Faced with a major new competitor intent on carving up the former two-portion pie three ways, their immediate reaction will be to cut fares. Eurotunnel anticipates that ferry fares will drop by around 10 per cent while flight fares will be cut by between 10 per cent and 25 per cent depending on the route. Eurotunnel is committed to offering competitive fares to match those available on air and sea routes, rather than trying to price their competition out of the market. The big advantage for the Tunnel operators is that their running costs are likely to be substantially lower than those of their competition. The business of getting people and goods through a tunnel is less labour-intensive than by air and sea. (But to counter that, they have huge loans to service.) In addition, it is less vulnerable than the other two modes to holdups caused by bad weather. Blue skies over the white cliffs of Dover or not, the Tunnel trains will still roll.

For the passengers the gains are enormous. The Tunnel will offer them a further alternative method of travel that will get them to most destinations faster. There will be greater access for UK foot passengers in the regions to the Tunnel rail network, through their British Rail station, and fewer holdups from switching modes of transport. It will be a reliable, all weather service, that does not need to be booked in advance.

Eurotunnel put a strong case for customs and immigration procedures to be carried out on trains during transit, rather than decanting passengers into a customs' area which takes valuable time and is inconvenient. It is more usual on Continental international trains and those passing over the Irish border for the formalities to be carried out on the train. For exporters the Tunnel offers the chance of cheaper transport costs making their produce more competitive in Continental markets. Currently over 60 per cent of all British exports head that way. If UK manufacturers can gain the competitive edge their market share will grow. This view is countered by research done at the Universtiy of Kent, which indicates that transport costs represent only a fraction of modern manufacturing costs. They are however more significant in the growth industries, distribution and tourism, both of which could benefit strongly from the Tunnel.

For all jobs that the Tunnel will create once operational, there will also be job losses. On the job gains' side, Eurotunnel estimates that 3160 new jobs will be generated in and around the Cheriton and Ashford terminals in 1993, rising to 3710 by the year 2003.

These workers will be employees of Eurotunnel and British Rail, involved in the management, operation and maintenance of the service and tunnels, as well as Customs and Excise and immigration control personnel. Their jobs and the wages they spend, will give rise to spin-off jobs at a higher rate than those generated during construction. Eurotunnel and the Kent Impact Study Team have set this at a multiplier level of 1.3 (it was 1.2 during construction) raising total job levels to 4100 in 1993 and 4800 in 2003. A further 550 jobs will be created at Waterloo's new international terminal.

But neither of these facts will be of much comfort to those currently employed in the east Kent ports of Dover, Folkestone and Ramsgate, many of whom stand to lose their jobs as a result of the Tunnel. Between 4300 and 6600 jobs are claimed as likely to be lost at the Kent ferry ports because of the fixed link. After subtracting the job gains for Kent and the negative effect on other employment which is dependent on port jobs, there is expected to be a net job loss of between 1400 and 4400 (Kent Impact Study Team). Over the succeeding ten years the job loss is expected to improve slightly, clawing back an extra 1700 to 3000 jobs by 2003.

Not surprisingly, the most vocal opponents of the scheme have been the ports of Dover and Calais and the ferry operators, banded together as Flexilink. They have been joined by the Transport and General Workers Union and the National Union of Seamen. Their argument is that the job losses will be much greater than either KEDB or Eurotunnel predict, ranging from 5000 to embrace all 40,000 jobs in the ferry industry. They argue that because short sea routes subsidise longer ones, the longer sea routes will be the first to be cut as fares on the more lucrative short crossings are pared to the bone to meet Tunnel competition. The result will be less choice for the consumer as the number of ferry destinations is reduced. Services that have been mentioned as likely to come under threat include those from Hull, Immingham, Great Yarmouth, Ipswich, Felixstowe, Newhaven, Portsmouth, Poole, Weymouth and Plymouth. The number of jobs likely to be lost was highlighted by the Transport and General Workers Union in its submission to the House of Lords during the passage of the Channel Tunnel Bill:

"The ferry industry nationwide and the ports and administration associated with it currently employ 37,000 people. That could be expected to grow to 50,000 by the year 2000 without a Channel Tunnel. All those jobs will be threatened by a project which will employ 5,000." (Transport and General Workers Union: February 1987.)

The TGWU has found itself in an awkward position over the issue of the fixed link. Because it represents a large number of ferry and port employees, it has taken an anti-Tunnel stance. But it also numbers among its ranks a sizable membership of construction workers for whom the Tunnel and its spin-off construction schemes mean a lot of jobs. While the main union was fighting to end the scheme, sections within it were busily negotiating the best deal they could for their construction members. Also it did no favour to port employees who saw the Tunnel as a potential saviour of Bristol, Liverpool and Clydeside.

Others with no particular political axe to grind have analysed the Tunnel's potential for generating growth at regional level, concluding that an infrastructure scheme on its own cannot boost the local economy without appropriate further associated investments. (R W Vickerman: *The Channel Tunnel and Regional Development: A Critique of an Infrastructure — led Growth Project* — March 1987). Mr Vickerman also suggests that the real magnets for industrial development lie 100 kilometres or so away from the Tunnel, around the M25 orbital motorway in Britain. (*The Channel Tunnel: Consequences for Regional Growth and Development*: June 1987). The first steps towards identifying exactly what appropriate investment could be attracted to Kent were carried out in 1986 when the Kent Impact Study was published. The work was done under the auspices of the Channel Tunnel Joint Consultative Committee, comprising representatives of UK government departments, Kent County Council, the local authorities affected directly by the scheme, Eurotunnel and British Rail, under the chairmanship of David Mitchell MP, Minister of State for Transport.

Its preliminary findings concentrated on an analysis of the current economic situation in Kent and suggested further areas of research to identify what the fixed link's impact would be. It spawned a second study, published in summer 1987, focusing on the need to maximise tourism and distribution and opportunities for diversification. It assessed the training and infrastructure improvements which the fixed link and other activities would require, and looked at alternative activities for the ferry ports and towns hit by the Tunnel's operation.

Dover Council, who initially opposed the Channel Tunnel scheme, has since adopted the pragmatic view that whether the Tunnel comes or not — "reduced employment in the coal mines and the new generation of ferries make diversification of the local economy imperative". Further, the Chief Executive has warned "The state of the town demands a tangible and rapid lead from this Council". (Dover District Council: Chief Executive's Report: *The Future of Dover District — A Forward Capital Programme*: 19 November 1986). The council is intent on promoting investment in tourism, warehousing and distribution and plans to spend £100 million up to the year 2003 in projects which will act as "pump primers" in attracting private investment to Dover.

Tourist schemes being proposed include a Roman Heritage Centre, town centre pedestrianisation, refurbishment of leisure centres and swimming pools and environmental improvements. New retail schemes are in the pipeline and the council is promoting the sale of prime sites in the town made vacant when the council relocated its offices, and industrial land is being provided for new industry.

For the county as a whole, Kent's track record in attracting new industry has been well below the average for the south east. But in 1986 it succeeded in bringing in its first Japanese company when Fuji Seal decided to establish its European headquarters and main manufacturing plant in Gillingham to produce specialist packaging. Further Japanese moves to the UK are under negotiation, in line with the current trend for Japanese industry to set up in the UK. In the 1980s Nippon Industry has moved into the UK in increasing strength in the high-tech industries in Scotland, Wales and the M4 corridor, and in the north-east with the Nissan car manufacturing plant in Washington.

In the competition between the UK and France and Belgium to attract new industry in the proximity of, or related to, the Tunnel, the UK has one big advantage. There is a tradition of USA and Japanese firms preferring the UK as a relocation base — the most oft-cited reasons being that we speak English and that a successul example is always worth following. Kent is also hoping to draw in newcomers from London companies choosing to decentralise. Better road and rail links should act as an incentive as should cheaper house and office prices. Kent may also have an inbuilt advantage, which the Nord-Pas de Calais region of France does not, of being closer to the nation's capital, and its huge markets, and therefore less disruptive for firms choosing to move out of London.

For UK ports the outlook is bleaker. They already face severe competition from Rotterdam as an international shipping base. Much of their traffic involves trans-shipment to Rotterdam, and with Eurotunnel and British Rail chasing the same business, UK ports could lose some of their most profitable trade. Long sea routes will also be affected as exporters take advantage of the economies of scale that ports like Rotterdam can offer in transporting the bulk of European exports.

The French ports of Le Havre and Dunkerque are expected to become more competitive in pursuit of a greater share of international trade, with considerable help from the French Government which has invested heavily in them over the last 20 years and is keen to see a return on that outlay. UK ports are likely to respond in one of two ways. If they fail to maintain their international market share, it may trigger internal competition as ports in the south vie with those further north for the crumbs that remain. North east ports could see their Scandinavian trade diminish as more efficient southern ports muscle in on what has traditionally been their market. The Container ports of Southampton (55 per cent freight) and Felixstowe (25 per cent freight) could lose trade to more efficient neighbours like Dover.

Alternatively, they could elect to streamline their operations and capitalise on their assets. Most ports have substantial landbanks which could be developed for distribution, industrial and leisure schemes. Currently Southampton is planning a major leisure scheme in its docks and at the other end of the country, Clyde is looking at a leisure and dock development scheme. If these succeed, they and others like them will generate substantial rents.

Unfortunately not all docks are able to use their landbanks to provide income. Many are barred by statute from doing so and , without a change in the law, will be forced to shuffle their feet on the sidelines while others compete for the spoils. Ports who want to streamline their operations are further handicapped by having rigid employment policies imposed which prevent them from shedding staff in the drive for greater efficiency. Some of these could close as a direct or indirect result of the Tunnel's opening.

Not all British ports have taken such a pessimistic view. In Scotland, the Clyde port is planning to take advantage of the opportunities that the fixed link will provide to develop as a deep water trans-shipment port, possibly on the basis that "if you can't beat them, join them". The Clyde is the first of the UK west coast ports to take on the challenge of competing with Rotterdam and Bremerhaven for North Atlantic business. Potentially it offers customers a 36 hour saving in sailing time from the east coast of America over Continental ports. It can also provide deep water berths, modern jetty facilities, a container port and a major dry bulk facility.

Although the idea of the Clyde taking trans-Atlantic traffic has been belittled by ports' organisations, the Scottish Office was sufficiently keen to fund a feasibility study into the scheme, in recognition of the considerable commercial spin-offs for Scotland.

If the rest of the UK were similarly far-sighted and resourceful, perhaps we would minimise the risk of becoming, what one Tunnel opponent called, ''a backwater in the European economy'' and instead ride the tide of Tunnel-induced economic growth.

7

La rêve de Napoleon... et al.!

Mick Hamer

The Channel Tunnel will be cancelled, warned the television news. It was Friday, 17 January 1975. The newspapers picked up in the story the next morning. The long-cherished hopes of building a tunnel to link France with Britain were once again facing relegation from the realm of realism back into fantasy.

The Tunnel had been in trouble for several months. Harold Wilson's Labour Government had already failed to ratify the Franco-British treaty by the deadline of 1 January 1975. It had been given an extra two week's grace.

The delay was not entirely the Government's fault. There had been two general elections in 1974, which had upset the parliamentary timetable for legislation. But the Government was also seriously worried about the escalating cost of the Tunnel, which some estimates were putting at between £1.5 and £2 billion. On the other side of the balance sheet, the Government was looking at a cancellation cost of about £20 million.

The Sunday newspapers were full of the dark end of the Tunnel [1]. A last minute plea from the French to complete the second phase of the works, which involved boring a 2.4 kilometre long pilot tunnel, was rejected.

On Monday, Members of Parliament were told of the UK Government's unilateral decision, a full two days after they had read it in the newspapers. In a three hour emergency debate the Government had a majority of 76.

The immediate cause of cancellation was that the private companies, involved in the project, had asked for the right to withdraw their money. The Government refused to countenance this demand. The companies for their part were worried about the British Government's evident lack of enthusiasm, and wanted the guarantees as a hedge against cancellation. But the underlying causes were far more important than this relatively trivial disagreement.

The recriminations started almost immediately. The Elysée, under the right-wing Government of Giscard d'Estaing, blamed the Labour left. Downing Street, in off-the-record briefings said the French Government was divided. The French Channel Tunnel Company was more to the point: "The British are a bloody nuisance"[3].

The Channel Tunnel Bill, which had already passed major parliamentary stages was allowed to lapse. Some 240 workers at the British end of the Tunnel were laid off. And a hole, about 350 metres long, under Shakespeare Cliff, near Dover, pointed out to sea in the general direction of France. The French also had a similar hole, pointing towards England.

A £500,000 tunnelling machine lay mothballed at the British end of the tunnel. The Department of the Environment's property services agency was delegated to look after it. The machine's story has all the tunnel's essential farcical elements, writ small. For the first 18 months the agency sent someone into the tunnel to tend the machine. This was stopped as an economy measure when a spare part was needed. Then water got into the tunnel and damaged the machine. It was subsequently sold for scrap. The contract price was £20,000. But the scrap merchant could not get all of the machine out of the tunnel. So the rusting edge of British technology still lies 350 metres out to sea.

The story of this attempt to tunnel the Channel begins some 20 years before, with a twice-tasted breakfast. Two sisters were seasick on a cross-Channel ferry. One was married to a French count and the other to an American lawyer. The result of this stomach-rending altercation with Poseidon was Technical Studies Inc, which was backed by the US bankers J Pierpoint Morgan, to look at the feasibility of a Channel Tunnel[4].

Events in Egypt, where President Nasser had nationalised the Suez Canal, also gave the prospects for a tunnel a decisive push. By the end of 1956 the Suez Canal Company suddenly found itself without its main asset. The Channel Tunnel Company was in a similar position — with the relatively trivial exception that its main asset had never existed. The

Work at Shakespeare Cliff near Dover in 1974 made use of earlier excavations, known as
the 1880s Beaumont Tunnel. The access tunnel to the foot of the cliffs was not excavated at
this stage although supports for its ramp were already in position.

company was a relic of an earlier attempt to tunnel the Channel some 80
years before.

Leo d'Erlanger was a life-long enthusiast for the Channel Tunnel. He
had been chairman of the Channel Tunnel Company since 1949, a
position that ran in the family, since both his uncle and grandfather had
also been chairmen. D'Erlanger started talks with the French Suez Canal
Company, which had independently been approached by Technical
Studies Inc. These groups came together in the Channel Tunnel Study
Group, which was set up in 1957 [5].

Almost immediately the military trotted out its objections to the Tunnel, as it had done for the previous century. General Sir Edward Spears, with a fine grasp of geography observed: "Such a tunnel would link this island to the Continent irrecoverably." He concluded: " Physical ties... would soon link our fate to that of our Continental neighbours." Frequently the military objection was no more than a cloak for little Englander sentiments. Viscount Montgomery was also among the objectors: "Why make things easier for our enemies?"

In fact the military objections no longer carried much weight with the Government. In 1955 Harold Macmillan, who was then Minister of Defence, had been asked if the Government had any strategic objections to the Tunnel. He replied: "Scarcely any remain"[6].

In March 1960 the Channel Tunnel Study Group presented its plans to the Government. This prompted a rival Channel Bridge Group to be set up in December 1960. This group was set up in France, and was composed of automobile and banking interests. It presented its plans to the Government in October 1961.

The French and British Governments set up a joint working group to look at the merits of the rival plans in November 1961. The Channel Tunnel Group's scheme was for a double bore railway tunnel, carrying both ordinary passenger trains and a special shuttle service across the Channel for cars and lorries. It was estimated to cost £160 million when the working group reported in September 1963 [7].

The Government's working group dismissed the bridge. "Of the three possible solutions to the problem of crossing the Channel which we have studied, the bridge is in our view the least satisfactory. The capital cost of the project (£298.5 million) is likely to be about twice as high as that of the tunnel, but the difference in cost is not justified by a sufficiently large foreseeable increase in traffic."

"The completed project would be a serious hazard and source of delay to shipping. It could not be undertaken in advance of international negotiations which might be prolonged and the result of which would be uncertain."

The third option which the group studied was to improve the ferries. It said that ferries would be more expensive in the long run, and wrote off the possibility of hovercraft as "not having been fully tested." The group concluded that the best option was the railway tunnel. "The bored tunnel project presented by the Tunnel Group is fully developed and consid- ered, and based on the application of proven techniques. Their latest proposed arrangements for ensuring passenger safety can be regarded as satisfactory...

"The rail tunnel does not present the same practical disadvantages as a bridge... from the economic point of view it constitutes a preferable solution to the continued use and development of established means of transport."[8].

The public reaction was enthusiastic. An opinion poll in the *Sunday Telegraph* found that seven out of ten people in Britain favoured building a tunnel, and only one in ten opposed it.

On 6 February 1964 Ernest Marples, the Minister of Transport, and his French counterpart issued a joint statement giving the Tunnel the go-ahead. They said it was "technically possible and that in economic terms it would represent a sound investment of the two countries' resources...

"The two governments have, therefore, decided to go ahead with this project. The next step will be to discuss further, in particular the legal and financial problems involved" [9].

In 1964–5 the two countries carried out a major geological survey of the Channel along the proposed route of the Tunnel. Seismic soundings tested the rock at 500 metre intervals, while 73 marine boreholes were drilled, and 6000 metres of cores extracted to test the rock for its strength and permeability.

The tests virtually ruled out two of the problems that the tunnellers might face. The faults found were almost all of no significance. The deepest trenches in the Channel bed still left more than 25 metres of sound chalk above the tunnel roof. Pumping tests, in which water is pumped into a borehole under pressure to measure the rate of leakage of the chalk were carried out at 54 boreholes, again without finding any major problems [10].

The tests showed that the deep Chalk Marl, through which the Tunnel would be built, ran all the way from Britain to France. Therefore major engineering problems, causing delays in construction and an increase in costs were unlikely.

When Marples announced the Tunnel it had been assumed that it would be built by private enterprise. The election of a Labour Government in 1964 changed all that. Labour was keen on public sector enterprise. De Gaulle's France favoured private enterprise. The compromise was for the tunnel to be built and financed privately, but with the governments guaranteeing around 70 to 90 per cent of the loans. The tunnel would then be handed over to a state-owned company which would operate it.

In February 1967 the Government asked for private interests to submit their proposals. Three British-based groups responded. The Channel Study Group's plan was backed by most of the British merchant banks and the American banks of Morgan Stanley and First Boston. An Anglo-French group was led by Hill Samuel and Warburgs had a third.

The Government asked the consortia to merge. It took time to achieve. But by 1970 a combined group was formed, the British Channel Tunnel

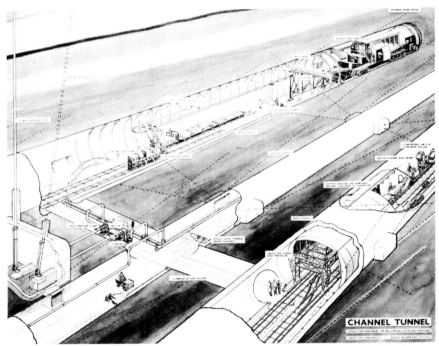

Layout of an earlier scheme. Many of the same features have been incorporated in the 1987 design and consultants Sir William Halcrow and Partners are still actively involved. But all dimension have been increased for the three parallel tunnels and cross-cuts, and it is no longer intended that any spoil be pumped to the seabed as a slurry.

company. It included the old Channel Tunnel Company, as well as British Rail, Rio Tinto Zinc and the banks. The French equivalent was the Société Français du Tunnel sous la Manche. It included la SNCF, French Railways.

By October 1970 the two companies and the Governments, prompted by Rio Tinto Zinc, had decided they needed further studies, at a cost of £1 million. The French felt that this was device to delay any tangible progress until after Britain had joined the Common Market. The British could then drop the tunnel.

There was deep distrust between the two countries. For in parallel to the two studies, the British Government commissioned another cost benefit analysis, which was kept secret from the French.

The revised proposals were accepted by the British and French Governments in March 1971. The two Governments and the companies agreed on a programme of still further economic and technical studies, which would "lead to a decision whether or not to proceed with the project."

Government, by its nature, takes time. Two Governments take even longer. So the formal agreement between the four parties was not signed until October 1972. This separated the work into three distinct phases.

The first phase was for initial studies, costing an estimated £5.5 million. These were to be completed by the middle of 1973. Half of the cost of these studies was to be raised by the companies, the other half under a Government guarantee.

The second phase of the work would be a trial bore of some two kilometres of tunnel by both Britain and France. If this was completed successfully, the project would go on to phase three, boring a complete tunnel. In March 1973 the Government published a green paper. It envisaged the formal agreement for phase three being signed in early 1975, continuing prophetically that only then would "the Governments and the companies... be fully committed to the completion of the project"[12].

So far the net result of 16 years of studying the Channel Tunnel had been to confirm that further studies were worthwhile. The phase one studies were finished in June 1973. The Government published its white paper in September. This approved the Tunnel "as an important and valuable link between the transport systems of the United Kingdom and of the continent." It said the Government would introduce a short Bill giving it powers for stage two of the work and a further Hybrid Bill for building of the main tunnel[13].

One of the most controversial parts of the white paper was a new high speed rail link between the Tunnel and London which would allow trains to run at up to 300 kilometres an hour and to take the larger loading trains that run in Europe. This possibility was barely hinted at in the green paper.

In Kent the phase one work had uncovered some weathered and friable chalk on the line of the Tunnel, where it crossed the coastline. So the route was shifted further towards Folkestone. In the White Paper this move was portrayed as shortening the tunnel route between England and France, which was economical with the truth, to the point of parsimony.

In France, the engineers opened up the old 1880s working at Sangatte, near Calais, pumped out the water, and dug some test bores. The British Channel Tunnel Company negotiated planning permission for its construction site with Kent County Council.

Meanwhile opposition to the Channel Tunnel was building up. There were two main sources. Environmental concerns about motorways disfiguring the landscape was at a peak. Kent County Council was keen on motorway building, as were all councils at the time. Even so, Kent's leader, John Grugeon, was moved to comment: "Kent does not want to become the lay-by of Europe."

The high speed rail link further incensed the environmentalists. The only high speed railway then in operation. It was very intrusive. British Rail said that the electric trains would not add significantly to noise levels, an assessment which *New Scientist* described as "fair". But even if the films are overstated the fears were real enough.

The other source of opposition was the Channel Tunnel Opposition Association, a coalition of ferry operators, such as Keith Wickenden's Townsend Thoresen and various road lobby groups. Sealink as part of British Rail, was not part of the opposition. However, Wickenden was a formidable opponent. He was a Conservative Member of Parliament. His ships, mostly called some variation on the theme of "free enterprise", carried the flag and his philosophy across the Channel.

There were also a couple of rival bridge proposals. The Channel Bridge Group wanted to build a series of suspension bridges, with seven spans of two kilometres and one of three kilometres. The bridge, 80 metres wide, would carry two railway tracks with two three-lane motorway carriageways on either side, on a box girder deck. The designer was a senior structural engineer at British Steel.

The other design, proposed by the consulting firm of Pell Frishman and partners, was for a two deck fully enclosed road and rail bridge. The top deck would carry an eight lane motorway. The bottom deck would carry extra traffic lanes for heavy vehicles and a twin track railway. The bridge was designed by Gordon Lorrimer, the Port of New York's chief architect. Almost exactly the same bridge was proposed by Eurobridge (and Pell Frishman) in 1983 [14].

The white paper dismissed these alternatives in just two paragraphs. Another idea that gained considerable support at the time was for a rail-only tunnel. The reasoning behind this was that it would produce a fast link from London to Paris without congesting Kent roads. This scheme was particularly popular with environmental groups. A leak [15] from the British Channel Tunnel Company showed that a rail tunnel would be economic. It would cost 30 per cent less to build and 85 per cent less to operate. However the option was not seriously considered, largely because work on the twin bore shuttle was too advanced.

Relations between both companies and the two Governments were still bedevilled by mistrust. In the early stages of planning the Tunnel consultants had produced forecasts of how much traffic it would attract. The British produced their forecasts, the French theirs.

In phase one it was agreed that the British and the French consultants should produce just one set of forecasts. This was important because sooner or later the companies would have to try and raise money on the basis of these forecasts. They stood scant chance of convincing the financiers if their own forecasters could not agree. So it was decided to split the work between the two teams of consultants. All the British parties then went away and secretly commissioned a complete set of

FRENCH TERMINAL

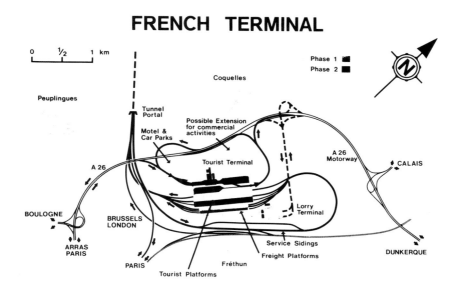

French terminal diagram. Plans in 1974 were not that different from those in 1987.

forecasts. The French did the same. So two sets of forecasts were again produced [16].

The British Government also commissioned two further studies. One was an attempt to dampen down environmental opposition by assessing the economic and social impact in Kent. It did not help the tunnellers. It was concluded that the tunnel would create an increase in employment of up to 6000 jobs, in the context of an anticipated growth of 96,000 jobs. The Government also commissioned a cost-benefit analysis. This was kept secret from the French, even though the same consultants that were working on the Tunnel with the French were commissioned. They lied to the French [17].

In October 1973, the UK Parliament debated the Channel Tunnel for six hours. A motion to study a rail only tunnel was defeated by 70 votes. The Transport Minister, John Peyton, defended the Government with a delightful mixture of pronouns: "in the present state of our affairs, where massive motor traffic is considered justified by all who use it and depend upon it, someone in my position is left with no choice but to try to accommodate it as decently and as tidily as one can" [18].

In November 1973, both Governments signed the Franco-British treaty which allowed the trial bores of phase two to begin. The treaty had to be ratified by January 1, 1975. The whole phase two was to cost £30.8 million. Ever enthusiatic, the French actually got to work two days before the treaty was signed. The British began the day after.

Despite the October vote, the British still needed to pass two major pieces of legislation. First the Channel Tunnel Bill and secondly, the Bill that would allow high speed railways to be built. In February 1974, the Tunnel received its first setback. In response to the miners' strike, Mr Edward Heath, the British Prime Minister, called a general election, and lost. Mr Harold Wilson became Prime Minister, leading a minority Labour Government.

In April 1974, the new Government reintroduced the Tunnel Bill. At Gravesend, on the River Thames, the engineering firm of Robert Priestly began to built the £250,000 tunnelling machine to dig the trial bore.

Conservationists continued to demand a public inquiry, to put the case for a rail-only tunnel, although this was a little like bolting the stable door after the horse had gone to the slaughter house. A loose and largely uncoordinated coalition began to oppose the Tunnel. The shipping interests, employers and unions were predictably opposed. In Kent, local councils called for the entire route of the high speed rail line through the county to be put in a tunnel. Anti-Europeans such as Member of Parliament, Mr Enoch Powell, became increasingly vocal in their opposition.

In London arguments raged over the site of the terminal for trains from Paris. British Rail wanted to build it at the White City. The Greater London Council (GLC) wanted Surrey Docks. Others favoured the traditional departure point for Europe, Victoria.

By July it was already beginning to appear that the Bill was too late to make the 1 January deadline for ratification of the treaty. In September the Bill was delayed yet again, when Wilson called another general electon. He was re-elected, this time with a majority.

By now the Government was having serious thoughts about going ahead with the Channel Tunnel. There was opposition to it from within the Labour Party, particularly from those, mostly on the left, who wanted Britain to leave the Common Market. Wilson warned Giscard d'Estaing in November that the tunnel might be cancelled. British asked France to delay the whole scheme by a year. France refused the request and in December ratified the treaty ahead of the deadline, to put pressure on Britain.

Britain missed the 1 January deadline. Three weeks later it formally cancelled the Tunnel. The cost of the cancellation, to Britain, turned out to be £17 million. For another £5 million the whole of the two kilometres pilot tunnel could have been completed [19]. Apart from the 350 metre long tunnel the only other tangible products of 18 years' work were 7,000 precast concrete tunnel liners, stacked by the Folkestone to Dover railway line, and a pile of consultants' reports, which in 1973 was estimated to be 60 centimetres high.

In its final form the cancelled Tunnel of 1975 was remarkably similar to the scheme which is currently on offer. It was a twin bore tunnel, with a smaller service tunnel, which would be used for maintenance or evacuation in case of emergency. The route was virtually the same. The fire precautions are also similar, with each of the ferry wagons having fire doors, or curtains, which would be fire resistant for 30 minutes.

If a train was evacuated the passengers would descend onto a platform running the length of the tunnel and walk into the service tunnel where ventilators would keep the air free of smoke. The British terminal for shuttle vehicles was to be at Cheriton. The main differences were the proposed London terminal and the high speed rail link.

Work on the British end of the Tunnel stopped quickly, ending on 25 January. The chief exception was, of course, yet another study. This time it was by a group of experts led by Sir Alec Cairncross. The Government had asked Cairncross at the beginning of 1974 to have another look at the Tunnel.

It was a curious study to have started, as Transport Minister Fred Mulley admitted: "many honourable members might think it odd — I would not argue too much if they did — that we are going through the process of enacting a bill to build the Tunnel...without having first decided whether to build the Tunnel" [20].

In the event the Tunnel was killed off half the way through Cairncross's enquiry, which naturally enough, turned into a post-mortem. The committee finally reported six months after cancellation. "We have been conscious, as our work proceeded that everything seemed to be happening in the wrong order. As our report bears witness we have been occupied in considering how a decision should be taken about the Channel Tunnel after a decision had already been taken. Studies on which we were asked to advise were too far advanced to be much influenced by any advice we could offer" [21].

However, its central conclusion was the same as every other study over the previous 18 years. "The Tunnel ... would produce an acceptable return."

Given the striking similarity between the 1975 Tunnel and the present project the reasons for cancellation of such schemes are sometimes instructive. But in 1975 it came down to a straightforward political choice, which has no modern echo, the Channel Tunnel or Concorde?

When it took office the Labour Government was faced with three major public sector projects, London's third airport at Maplin, the Channel Tunnel and Concorde. All three of these projects had stirred up sizeable public opposition. And the grand designs of the 1960s were going out of fashion. The Greater London Council, for example, had plans for a vast system of ringways, orbital motorways in London, costing £2000 million -

SECTION THROUGH CROSS ADIT

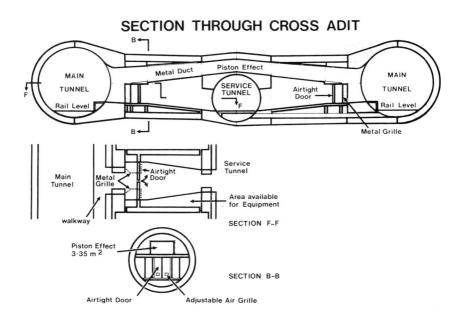

Cross-section of a cross adit in the 1974 scheme. In 1987 these are called cross-passages, occur every 375 metres and tunnel diameters have been enlarged at considerable cost to incorporate extra safety measures, particularly during construction.

roughly four times the cost of the Channel Tunnel. Labour gained control of the council in 1973 and in response to public hostility, axed the ringways.

In Government, Labour had no choice but to cut two out of the three major projects. Maplin went first. Tony Crosland, speaking privately in 1975, said it came to a straight choice in Cabinet committee between Concorde and the Channel Tunnel. He added that the major factor swinging the choice Concorde's way was the advocacy of Tony Benn. Benn was a Bristol MP and Concorde was made locally [22].

An additional factor was that because of Britain's unhappy experience with Concorde, the Channel Tunnel treaty had get-out clauses written into it. This made the Tunnel the soft option.

However, this suggestion of crude political choice neglects other underlying factors which told heavily against the Tunnel. Both Concorde and the Channel Tunnel were the subject of considerable environmental opposition, in the case of the Tunnel, skilfully orchestrated by the Defenders of Kent.

The opposition to Concorde was easy to understand. The aircraft was extremely noisy. The environmental case against the Tunnel was less clear cut. The Tunnel would cause considerable damage at Cheriton, although this was not large compared with other contemporary motor-way schemes, such as the Aire Valley, in Yorkshire, or the effect on Epping Forest of the M25. On the other hand the Tunnel had the environmental advantage of cutting the number of juggernauts on the roads of Kent.

The environmental case against the Tunnel was strengthened by the oil crisis of 1973–4, when oil prices doubled in under a year. It was a severe blow to the Tunellers' confidence. The environmentalists argued that the traffic forecasts assumed that cheap oil would continue into the future. They argued that the forecasts ought to be revised downwards to take account of higher oil prices. The oil crisis jolted most Western Govern-ments into reviewing their public spending plans.

It was a measure both of the Government's weak commitment to the Tunnel and of the civil service attitudes in the main sponsoring department, the Department of the Environment [23], that the oil price rise affected the decision on the Tunnel. The motorway programme, which was then running at about one and a half Channel Tunnels a year, emerged almost unscathed.

The proposed high speed rail link also fomented opposition to the Tunnel. It had been added on at a late stage. The 1973 green paper had simply pointed out the difficulties for rail traffic between the Tunnel and London. British Rail's southern region is electrified on the third rail system. Main line trains in France (and Britain) are electrified with overhead wires. Britain's loading gauge, that is the clearance for trains when passing each other or passing under bridges, is smaller than in France. In addition there was a capacity problem, of finding time for the Channel Tunnel trains amongst the region's busy commuter services. The green paper discussed a number of options for dealing with these problems. The most radical was to enlarge four existing tunnels on the route to take the larger continental trains.

The white paper six months later baldly stated that a high speed rail link between London and the Tunnel was "essential". The decision was hasty and ill-conceived. It was the result of an internal wrangle on the railways, between British Rail's central management and its southern region.

Fierce rivalries still persisted on the railways, even 30 years after nationalisation. The southern region was the heir of the former Southern Railway. It still acted as though it was independent. It even had its own train designs, 20 years after the rest of British Rail had been standardised. During 1973 much of its activity was devoted to fighting off the encroachment of the Channel Tunnel on its metals [24].

The original set of improvements to the main line to London would have cost £125 million. This estimate involved the upgrading of southern region lines. The region resisted these plans, arguing that they would cause unacceptable disruption to commuter traffic.

In the face of this intransigence British Rail opted for an entirely new line, so that Channel Tunnel passengers would not have to mix with the commuters. And since there was going to be a new line it might as well be designed for high speeds. The costs escalated to £370 million — at which point it was almost as expensive as the Tunnel itself.

The high speed rail link became the most vulnerable part of the Tunnel, both because of the environmental opposition which it aroused and because of its cost. The British Government had to spend £234 million on building the Tunnel, half the total cost, and 40 per cent more for the rail link. Furthermore the cost of the rail link was a substantial underestimate. British Rail wanted another £40 million for freight services. And the figures excluded a large, and undefined, sum for compensating land-owners. The tail had begun to wag the dog.

Part of the responsibility lay with British Rail's board. It felt that the Channel Tunnel was peripheral. The chairman was Mr Richard Marsh, a former Minister of Transport. Later British Rail's director in charge of the Channel Tunnel, Mr Peter Keen, described Marsh as "of an emphatic and shrewd turn of mind, whose view was that the project was unlikely to happen and he was not going to help it happen" [25].

In 1974 Labour tried to cut the cost of the high speed rail line. Tony Crossland rejected the plan, and sent British Rail away to come up with more modest proposals. But by then the damage had been done. British Rail did come up with cheaper plans — and "intermediate" option. But it was too little, and above all, too late.

Cairncross produced the final verdict. Far from being essential, as the white paper claimed, Cairncross found: "the return on the additional investment in any of the rail links so far proposed is unlikely to produce an adequate return." The Tunnel was worth building. The railway was not [26].

Another element was the complete disarray of the tunnellers' management. On its most basic level the structure of management was bound to be complicated. Two Governments and two companies were involved. But there was no single project management team, although Rio Tinto Zinc did to some extent fill the part in the early 1970s. British Rail did not have a project management team for the high speed rail link, and was not included in the Tunnel management. The result was that with no central leadership the project drifted.

Naturally enough the British counted the cost of the Tunnel in pounds sterling, while the French counted in francs. This produced the problem that fluctuations in the exchange rate between the franc and the pound might affect the future cost and profitability of the Tunnel. So the

Four into one. The two tunnels on the left are British Rail tunnels through Shakespeare Cliff. The square adit in the centre goes down into the 1974 workings, turning a sharp right to travel seawards. The tunnel in the the the top right-hand corner was excavated in 1974 for access to the working platform and the Tunnel access.

consultants clearly had to predict the movement of exchange rates. The British Government said that the consultants could assume any change they liked, as long as they did not predict a devaluation of the pound. The French adopted the parallel pose. The consultants could assume any change they liked as long as the franc did not go down against the pound.

An additional problem was that most of the British involved did not speak French well. Their greatest triumph was to understand what had been said, rather than to argue a point. The result of this basic lack of understanding was that meetings could end in shout-ups. On one occasion the French stalked out: "Messiers les anglais. Si vous voulez la guerre, vous aurez la guerre" [27].

The final flaw was that Britain and France have two very different political systems. The environmental lobby had great influence over delaying, and defeating the Tunnel, which the French found difficult to understand. In one story a French official was asked "What would happen if the French parliament turned it down?" He replied: "They would not. And if they did we would ignore them."

With the cancellation of the Tunnel came the end of d'Erlanger's dream. Now in his middle 70s he would never see the Tunnel open. But he was determined to be on the first train. So he charged an executor to carry his coffin through on the first train [28].

After 1975 it was close season for the Tunnel. But by 1978 interest in the Tunnel was beginning to increase again. British Rail now recognised that it had lost an opportunity. And in Sir Peter Parker, the railway had a chairman who was determined to fight for a better deal for the railway — and one who was not prepared to tolerate secession by the regions.

In September 1978 Parker announced that British Rail and la SNCF were working on a plan for a single bore railway tunnel, coupled with a service tunnel. The cost was (then) around £700 million. "Roughly the price of a score of jumbo jet aircraft. Jumbos last about a decade and a half. Tunnels are more or less for ever" [29].

British Rail did not publish its plan until April 1979, just before the May election. The Conservatives won the election, under a Prime Minister who steadfastly refused to travel by train.

British Rail's plan was for a six metre diameter tunnel. Trains would run through the Tunnel in flights, with half a dozen leaving England for France, and then half a dozen returning, once the last train had left the Tunnel.

British Rail set up a consortium, including the civil engineers Costain, the French company Spie Batignolles, and two merchant banks to build and finance the Tunnel. In 1980, Transport Secretary Norman Fowler said that the Tunnel had to be privately financed.

The British Government set up another independent study to look at the rival schemes, selecting Cairncross for the job. Cairncross looked at the range of alternatives, from a single bore railway tunnel, which had become known as "the mousehole", to a bridge. He concluded that the best bet was a twin bore rail tunnel, with a shuttle for cars and other vehicles. This was the scheme being promoted by the Channel Tunnel Development Group, which included Tarmac and Wimpey. A similar scheme was being promoted by another consortium of Balfour Beatty and Taylor Woodrow.

British Rail dropped out and the five British civil engineers (Balfour Beatty, Costain, Tarmac, Taylor Woodrow and Wimpey) merged to form the Channel Tunnel Group. At the beginning of 1981 Kenneth Clarke, the junior Transport Minister, predicted that there would be a final decision on the Channel Tunnel by the end of the year, saying "We hope to reach a provisional decision by the end of the year as to which scheme might go ahead." It took another five years to achieve that goal [30].

In September the new French President Francois Mitterrand had his first meeting with Margaret Thatcher, the British Prime Minister. The officials were desperate for an issue that the two leaders, from socialist France and right wing Britain, could agree on. It was the turn of the Channel Tunnel.

Plans for reversing the melting of the last ice age, which destroyed the physical link some 8000 years ago, and reuniting Britain with the continent go back nearly 250 years. In 1751 Nicolas Desmaret, a farmer, wrote a "dissertation sur l'ancienne jonction de l'Angleterre a la France". He observed that the wolves in Britain must have crossed from the continent when there was a land link, because they could not have swum the Channel. He argued that this "communication d'origine divin" should be restored, either by a bridge, a tunnel or a causeway [31].

The first reasonably serious proposal for a link came in 1802. Britain and France had just signed the Treaty of Amiens, and peace broke out. A French mining engineer, Albert Mathieu-Favier interested Napoleon Bonaparte, the first consul, in building a tunnel between the two countries to symbolise the new era of peace.

Napoleon was impressed by the plan and discussed it with the Whig statesman Charles James Fox, who responded "this is one of the great enterprises we can now undertake together." In 1803 Tessier du Mottray came up with an alternative plan for a submerged tube tunnel. Marginally more practical was the proposal by the Montgolfier brothers for a series of hot air balloons, each carrying 3,000 travellers, to ferry across the Channel. None of these enterprises got any further, because by the end of 1803 relations between Britain and France were back to normal. The two countries were at war [32].

The war probably cost fewer lives than Mathieu-Favier's tunnel would have done. He wanted to build a bored tunnel. The traffic would be horse-drawn. The tunnel would be lit by candles and ventilated by flues, reaching through the sea into the fresh air. On the Varne sandbank Mathieu-Favier planned an artificial island for the horses (and passengers) to regain their breath. This was a fairly essential detail. For on the five hour journey through the tunnel the combination of burning tallow and horse manure would have been an overwhelming olfactory sensation.

The Napoleonic wars and the subsequent cool relations between England and France caused a hiatus in tunnel planning. The renewal of interest in the 1830s reflected not just an improvement in relations between Britain and France — by 1830 the two countries had been at peace for 15 years — but also the result of technological change.

One technical advance was the coming of the railways. In 1830 the Liverpool and Manchester Railway had opened, the first steam-operated passenger railway. In 1825 Marc Brunel had begun tunnelling under an altogether more modest stretch of water, the River Thames. Brunel used his own patented shield. This was a casing which protected miners at the tunnel face until timber supports could be erected. The shield was pushed forward as tunnelling progressed [33].

Brunel's tunnel, between Rotherhithe and Wapping, in east London, took 16 years to finish. It was beset by liquidity problems, both financial and technical — through flooding. However, his example inspired tunnellers to new depths.

Aimé Thome de Gamond was a man of considerable talent and fortune. He expended both on the Channel Tunnel. He was a doctor, lawyer and

The age of steam has gone but its markers, as ventilation tubes for the Dover to Folkestone railway line, extend along the cliff tops near the village of Aycliffe. When these stopped being classed as 'eyesores' they suddenly became 'industrial archaeology' and in need of restoration.

engineer. In 1833, at the age of 26, he carried out a hydrographic survey of the sea bed between Britain and France. The next year he produced the first in a long line of plans to link Britain and France. His first idea was for a submerged tube. He would level the sea bed, with a shipboard rake, and then lay a cast iron tube, which would be lined with brick [34].

In 1835 he proposed a submerged arch. The arch would be built with a shield which would move across the sea bed. It would take just 30 years to build. The following year he proposed a bridge, and sketched out four different designs. The 1837 plan was for a train ferry. Two moles eight kilometres long would project from Britain and France. A steam ferry would ply in between. In 1840 he wanted to build a causeway, by dumping blocks of concrete on the sea bed, leaving three openings for navigation. It was abandoned, he said, "because of the obstinate resistance of mariners."

It was the 19th century equivalent of Trivial Pursuits, a game anyone could play, although not necessarily win. Some of these plans were decidedly sketchy. In 1842 Mr E Pearse proposed a submerged tube, in a letter to the Railway Times. A similar scheme was put forward by Mr De la Haye, who despite his name, came from Liverpool. His tube, in 130 metre lengths, would be bolted up undersea, by workers without diving equipment. Ferdinand's 1848 plan was for a floating tunnel, secured by buoys.

Technology, finance and commonsense provided no bounds for Victorian imagination. Hector Horeau had no lack of the latter. His 1851 scheme was for a submerged tube. The ventilation shafts would be decorated by gothic turrets. Two lighthouses would mark the ends of the tunnel. Glass lights, let into the roof, would provide illumination and the railway carriages would have glass roofs.

The operation of this tunnel depended on gravity. Horeau believed that the Channel was at its deepest in the middle, near the Varne sandbank. The carriages would be propelled downhill by their own weight, and some of the way up the other side, where compressed air propulsion would take over. The cost of this scheme was £87 million.

Prosper Payerne was a doctor from Grenoble. In 1842 he had won a challenge from the East and West India Company, by staying under water in a diving bell for 12 hours off the coast at Cherbourg. He later repeated the trick, somewhat more spectacularly, in Paris, by having himself and the diving bell thrown into the River Seine.

It was this wealth of experience which the doctor brought to bear on the problem of the Channel. His 1855 plan involved the construction of a causeway across the bed of the sea, and then creating an archway with prefabricated blocks, using 40 subaqueous boats.

James Wylson's 1855 plan for a submerged tube would have cost £15 million. It was proposed in a letter to the Illustrated London News. In the same year Favre proposed a bored tunnel. James Chalmers' contribution, in 1861, was a book, advocating a submerged tube. Rather more inventive was the Abbe Angelini's idea for dropping a tube on to the sea bed. It would then work its way down, under its own weight, to the chalk that lay below, some 58 metres down. If the game had a prize for originality then Doctor Lacomme would be a leading contender. He wanted to dispense with a tunnel entirely, and run his submarine railway trains directly along the sea bed.

Meanwhile de Gamond was allowing his ideas to become tinged with realism. In 1855 he carried out a geological survey by diving down to the sea bed and collecting samples. It was a considerable feat for a man in his late 40s, to dive 30 metres without breathing equipment. On one occasion he was attacked by eels which he described as "malevolent fish, which seized me by the legs and arms".

De Gamond's 1856 scheme — he seems to have produced at least one a year — included an international port on the Varne sandbank. Two tunnels would connect it to Britain and France. Spiralling railway tracks would link the sandbank to the tunnels, in much the same manner as they linked the EuroRoute scheme of the 1980s.

The scheme was presented to the Government of France in 1857. The Emperor, Napoleon III, was keen on the idea, and referred it to a commission of French scientists. In Britain some of the leading engineers, including Isambard Kingdom Brunel and Robert Stephenson, backed the scheme. Prince Albert also approved — no doubt because Queen Victoria suffered badly from seasickness. Lord Palmerston, the British Prime Minister, did not: "What! You pretend to ask us to contribute to a work the object of which is to shorten the distance which we find already too short."

Hedging his bets de Gamond produced an alternative suggestion for a bridge between Calais, and East Ness. Gustave Robert wanted a causeway, wide enough for four railway tracks. However, the attempted assassination of Napoleon in 1858 (the bomb was said to have been made in Britain) killed the Tunnel, despite the recommendation of the commission that £20,000 should be spent on exploratory works.

In 1865 de Gamond tried again, this time in association with two British engineers. One was William Low, who had helped to design Box Tunnel, on the Great Western Railway. They carried out more geological studies and in 1867 they presented their scheme to Napoleon III and the British Government. Queen Victoria said: "You may tell the French engineer that if he can accomplish it I will give him my blessing in my own name and in the name of all the ladies of England."

Although de Gamond's name was attached to the plan, he was now old. He wrote to Low handing over the spritual responsibilty for the tunnel. "I have always hoped to live long enough to become the coadjutor

Engraving published in 1803 illustrates fears of invasion, with or without a tunnel, which remained well into the twentieth century.

of an English engineer.... Providence has willed that you should be that man.

"The direction of the work belongs to you legitimately. On my part I will give you the mental aid of my old experience" [35].

The joint committee of British and French scientists, set up in 1868, approved the plan in 1869. The Board of Trade, in Britain, said there would be no Government money for the tunnel.

Almost every scheme that has ever been put forward to cross the Channel has quickly produced a rival. The late 1860s produced a plethora of plans. John Bateman and Julian Revy wanted to build a submerged tube in 1869. The tubing, which was four inches thick and cast in three metre long sections would be bolted up by workers in diving bells on the sea bed. Bateman and Revy considered the tube so stout that no protection from anchors was necessary. Napoleon said: "C'est le seul realisable" [36].

A Channel Bridge Company was set up. The magazine *Nature* complained that it had produced no definite plans. [37] A 1868 plan had envisaged "ten spans each over 9,000 feet". The rumour was that the company had increased the number of piers to 30, and cut the length of the spans to 3000 feet (slightly under 1000 metres). Only a handful of bridges exceed this span today. The magazine remarked sarcastically: "the only thing wanted to complete this great national work in *three* years appears to be a subscription of eight million sterling to the credit account of the Channel Bridge Company."

Sir John Hawkshaw, an English civil engineer devised the chief rival to de Gamond and Low. Hawkshaw had already built the railway tunnel under the Severn, as well as the roofs of the railway terminii at Cannon Street and Charing Cross — the latter part collapsed disastrously in 1905. He had carried out some test boring at St Margaret's Bay, to the north east of Dover between 1865 and 1867. Hawkshaw favoured a single bore tunnel, to carry two railway tracks, in contrast to Low who supported a twin bore tunnel.

The committee of French and British scientists reported favourably: "driving a submarine tunnel in the lower part of this chalk is an undertaking which presents reasonable chances of success". The Empress Eugenie, returning from opening the Suez Canal in 1869, succeeded in interesting its engineer, Ferdinand de Lesseps, in the Tunnel. But the Franco-Prussian war of 1870 led to the collapse of the second empire. Napoleon and Eugenie fled to England, by boat, settling in Chislehurst.

The war also finally defeated de Gamond. His fortune gone, he ended his life in straightened circumstances, supported by his daughter, who gave piano lessons. He died in 1876.

The war caused the tunnellers some pause. Low had cooperated with Hawkshaw, who was very much the senior of the two, in about 1868. But by 1872 the two had parted company. Hawkshaw clearly had more pull with the establishemnt. Low had the better technical scheme.

The chief advantage of Low's scheme was cost, which is the reason that subsequent bored tunnels have opted for Low's twin bores, rather than a single bore. The cost of a tunnel is roughly related to the square of the tunnel's diameter. If one tunnel is twice the diameter of another, it costs about four times as much to build. So two small tunnels are much cheaper to build than one large one.

Low's twin tunnels also had the advantage that they provided an escape route in the event of an accident. The only important defect in Low's scheme was his intention to use compressed air locomotives to haul trains. Compressed air had been tried out on the North Metropolitan Tramway in north London, unsuccessfully. And trams in Nantes ran on compressed air until as late as 1913. But it was scarcely practical for the Channel Tunnel. Low's scheme also had the operational disadvantage of two large hydraulic lifts, similar to those used on canals, which would lower passengers and trains into the tunnel.

Hawkshaw planned to operate his tunnel with steam engines. The direct connection to the French and British railways was a distinct advantage, but the ventilation problems were not inconsiderable . Hawkshaw intended to use steam engines which consumed their own smoke. Similar engines were in use on the London underground, called Fowler's ghosts by the locals, after the designer, Sir John Fowler. Conditions were so bad on the "drain", as it was known, in the early 1870s that the company was forced to cut extra ventilation holes in the road between Edgware Road and King's Cross — a solution that was not entirely applicable to the Channel Tunnel.

Hawkshaw and his backers set up the Channel Tunnel Company in 1872, although an informal committee had existed for the previous four years. The company persuaded the Board of Trade to allow further trial digging at St Margaret's Bay.

In 1873 Low set up the rival Anglo-French Submarine Railway Company to pursue his plans, but failed to raise any finance. He was reduced to appearing at a public enquiry in France to oppose Hawkshaw's plans.

In 1875 a Channel Tunnel Bill was submitted by the Channel Tunnel Company, giving its powers to build a pilot tunnel. The bill's preamble says that the Channel Tunnel is the next most important project after the Suez Canal. The bill was passed on 2 August 1875. France also passed similar legislation on the same day, giving the corresponding French company headed by Michel Chevalier, the chief of mines, a 99 year lease on a site at Sangatte, near Calais.

The French legislation went into great detail about the fares and the way in which the tunnel would be operated. The first class fare was to be 50 centimes per kilometre. On the other hand a coffin and a corpse was to be charged one franc 50 centimes, thus opening the way for spectacular legal cases over passengers dying in transit — a not altogether unlikely event given Hawkshaw's ideas on ventilation.

On 30 May 1876 the two countries signed a protocol laying down the basis of a treaty, governing the construction of the tunnel. The treaty had to be ratified within eight years of 2 August 1875. The tunnel had to be

An immersed tube scheme was mooted many times, with this design proposed by M. Hector Horeau around 1850. The first commercial immersed tube crossing in Britain is under construction across the Conwy estuary in north Wales in 1987.

working within 30 years, or the Governments could award the concession to another company. In return the companies gained a 99-year monopoly.

The French began work enthusiastically, starting a tradition that continues until the present day. The engineers took 7700 soundings on the line of the route, and collected nearly 3300 samples from the sea bed. In 1878 L'Association du Chemin de Fer Sous-marine entre La France et Angleterre sunk a shaft at Sangatte and began digging towards England. However, there was scarcely any activity around the shed of the Channel Tunnel Company in St Margaret's Bay. The company could not raise the money to start the scheme.

So in the middle 1870s, Low went to Sir Edward Watkin to solicit support for his tunnel. Watkin was chairman of the South Eastern Railway and the Manchester, Sheffield and Lincolnshire Railway. He was an outsize railway baron who dreamed of linking the Manchester, Sheffield and Lincolnshire Railway to Europe. He did eventually get as far as London, when the railway was renamed the Great Central. The London extension was built to the larger continental loading gauge, at considerable expense, especially to take cross-Channel trains. The Great Central main line was closed during the Beeching cuts of the 1960s, just as Britain and France began to talk seriously about the tunnel again.

Low had made an astute choice. Hawkshaw's company was tied up with the London Chatham and Dover Railway (London smash'em and turnover, as it was popularly known, on account of its accident record). This railway's tracks passed the almost unused digging site at St Margaret's Bay.

The South Eastern Railway (slow, easy and comfortable) and the London Chatham and Dover were rivals. Low had always favoured a more westerly route, on geological grounds. The South Eastern's tracks approached Dover from the west. One of the objections that was later made by the Hawkshaw camp to Low's route was that the tunnel could not "be connected with the London, Chatham and Dover Railway Company's line at Dover". That was precisely Watkin's intention.

In 1880 Watkin sunk his first shaft at Abbot's Cliff, and dug a pilot tunnel for some 800 metres of so. In 1881 he floated the Submarine Continental Railway Company, controlled by the South Eastern Railway, to further Low's tunnel. The company had £250,000 capital and under the South Eastern Railway Act of 1881 it gained the legal power to dig a pilot tunnel — a small matter which Watkin's works of 1880 seem to have overlooked.

This time Watkin began work on a second pilot tunnel at Shakespeare Cliff. He reached an understanding with the French company, which continued digging towards England with renewed enthusiasm. The

French used the conventional tunnelling technique of the time, drilling and blasting. It risked causing a fracture in the rock and a leak. Watkin turned to modern technology, in the form of a compressed air tunnelling machine, designed by Colonel Frederick Beaumont.

By 1882 the work was progressing well. Watkin's forte was public relations, and a succession of parties was held to impress politicians, financiers and other persons of influence. Sometimes the meals were held in the tunnel itself, which was illuminated by electric lights. In February Watkin forecast that the first mile of tunnel would be finished by Easter. In March the reporter of the Illustrated London News, in the fine and sober tradition of the press was clearly impressed by being given two lunches, one in the tunnel and one at the Lord Warden Hotel. But as the tunnel began to look more realistic then so the military became more concerned [38].

High above the cliffs of Dover in the castle, Wolseley had visions of hordes of Frenchmen emerging from the tunnel mouth in the dead of night to murder the Dover burghers in their beds. "The proposals to make a tunnel under the Channel, may, I think be fairly described as a measure intended to annihilate all the advantages we have hitherto enjoyed from the existence of the silver streak...."

The *Sunday Times* agreed. It warned that the existence of the Channel prevented the movement of "nihilists, internationalists and Bradlaughites". (Charles Bradlaugh was a radical member of Parliament, atheist and campaigner for birth control.)

Public opinion was now moving against the Tunnel. Queen Victoria found it "objectionable". *The Times* newspaper declared: "the silver streak is our safety" and a letter writer warned that a French cavalry force might be sent through the tunnel in three hours and secure the English approach — always assuming it escaped being mown down by trains on the way.

Even those who were in favour went on the defensive. William Gladstone later wrote: "national opinion was too strong against it... it was too strong for me." Watkin's public relation techniques did help to still some opposition. One opponent is reported to have said: "I am strongly against the construction of the tunnel, and I told Watkin so. But he gave...an excellent luncheon so I should not like to do an unhandsome thing to him by signing the protest."

The Board of Trade called a halt to the digging at the beginning of April 1882 on a technicality. It said that the land between the foreshore and the three mile limit was Crown property. Any tunnelling would involve trespass. Watkin was a man of considerable determination. He continued tunnelling.

He wrote to the Board of Trade saying that the boring machine had to be kept on, on the grounds of safety. The compressed air machines ventilated the tunnel, and that switching them off endangered life. Joseph Chamberlain, the President of the Board of Trade, said the machine could be kept on if life was endangered. He tried to send an inspector, Colonel Yolland, to see for himself. Watkin then prevaricated. First he was out of town when the inspector wanted to visit. Then repairs to the machinery made the tunnel unsafe. Next a breakdown of the winding gear made the visit impossible.

All the same Watkin claimed that he had bought ancient manorial rights, which allowed him to dig out to the three mile limit. The Board of Trade asked him for his title, which Watkin seemed unable to produce.

The delays continued into May and June, and still Yolland could not inspect the tunnel. By now the Board of Trade suspected that Watkin was still tunnelling. From the cliff top its officials had observed buckets of spoil being removed from the tunnel. The Board threatened legal action, and on 5 July obtained a court order forcing Watkin to stop, and to allow an inspection.

Even then Watkin still had a trick. The inspector arrived to find that the company was not going to provide him with the tools he needed. He was forced to return to London and bring his own surveying equipment. He finally examined the tunnel on 15 July. Three weeks later he paid another visit, his tape measure revealed that the tunnel was now 70 feet longer. Watkin narrowly avoided being punished for contempt of court by arguing that Yolland had made a mistake in his figures.

A joint committee of the Houses of Commons and Lords was set up to consider the question of a Tunnel. An angry mob smashed the windows of the Channel Tunnel Company's offices in London, which was a mite unjust, since Hawkshaw's company had nothing to do with Watkin's tunnel.

The committee's hearings were less about how the Tunnel might safely be built, and more to do with how it might be blown up. In 1883 it decided, on a six to four vote, that the demolition techniques were unreliable for the Tunnel to be built. The French finally gave up hope in March, and stopped digging their tunnel. The French had dug 1800 metres, slightly less than the 2100 yards of Watkin's tunnel.

In 1886 the Submarine Continental Railway Company bought up Hawkshaw's defunct Channel Tunnel Company, and took its name. The next year Watkin tried again. But his bill was defeated by 76 votes in the House of Commons. Lord Derby commented that the "predjudice and panic are too strong to allow of any present attempt to construct a tunnel." Three more times in as many years new Bills were defeated.

A bridge scheme suggested by M. Hersent and M. Schneider in the late 1880s, for forging the cross-Channel link.

After 1890 Watkin gave up. Beaumont's machines were turned around landwards to dig for coal - at the South Eastern Railway's expense. However, Watkin still cherished his continental ambitions. In 1890 he began to build a vast cast iron tower on the fields outside London; at a place called Wembley. It was to be 50 metres taller than Eiffel's tower in Paris. Visitors would travel from all over London and up the Grand Central out to Wembley to visit the tower. It got as far as the first story, above the tree line. And there it stopped, because there were no visitors. It became known as Watkin's folly. Watkin died in 1901, the year the extension of the Great Central was finally completed with the opening of Marylebone Station. The folly was demolished in 1907. Its site is now Wembley Stadium.

Despite the political opposition engineers and eccentrics alike continued to dream up new ways of crossing the Channel. In 1889 two engineers from the French steel firm of Le Creusot, produced a plan for a bridge. This gained the support of Sir John Fowler, the English engineer who had suggested a train ferry in the 1860s as an alternative to the tunnel. In 1892 plans for a transporter were published. This would carry four trains at a time across the Channel. It would run on submerged rails 15 metres below the surface of the sea.

The signing of the Entente Cordiale, in 1904, was the signal to dust off the tunnel plans afresh. Albert Sartiaux, for the Société Concessionaire in France and the Channel Tunnel Company's engineer, Francis Fox, worked up another scheme. This followed Low's basic idea, of two twin tunnels, connected by cross passages to aid ventilation. This time the engineers proposed having electric trains, which greatly eased the engineering problems. They also planned a third drainage tunnel, which would be used to dig the cross passage, so that work on the main tunnels could proceed at more that one point to speed up construction. A narrow gauge railway in the third tunnel would remove the spoil.

They also included a special viaduct around the French coast, so that if the Entente broke down, the royal Navy could have a free pot shot to disable the tunnel. The British premier, Henry Campbell Bannerman, a Liberal, asked the Channel Tunnel Company for the time to consider. In 1907 the company, which since Watkin's death had been chaired by the banker, Baron Emile d'Erlanger, introduced a bill. But it was forced to withdraw. The vulnerable viaduct had failed to still the military objection. Campbell Bannerman argued: "Even supposing the military damages involved were amply guarded against, there would exist throughout the country a feeling of insecurity which might lead to a constant demand for increased expenditure, naval and military, and a constant rush of unrest and possible alarm, which, however unfounded, could be most injurious in its effect…"

James Kier Hardie, the chairman of the independant Labour Party had the military's measure: "I am a strong supporter of the scheme and consider the military opposition to it to be based on neurotic foolishness."

Despite strong military objections the consensus after the war tended to the view that the tunnel would have helped the war against Germany. In 1922 Maréchal Foch said: "If the Channel Tunnel had been built it might have shortened its duration by one half." The French premier, Georges Clemanceau and his British counterpart, Bonar Law, concurred. It was scarcely original thinking. In the 19th century the German Field Marshall, Helmut von Moltke said: "The tunnel should not be made, as it would not serve for invading England, but would be fatal to Germany in the case of a conflict…".

Immediately after the war the Channel Tunnel Company had received some encouragement from David Lloyd George. He told a deputation of members of Parliament that the 1907 decision was wrong. Thus emboldened the company decided to test a new tunnelling machine, designed for undermining trenches during the war, by the Leicester engineer Mr D Whitaker and refined in 1921.

The machine had an electric motor, and could excavate 1.3 metres an hour, a substantial improvement on Beaumont's machine. The trials were not altogether successful. On one occassion the machine's cutting heads became jammed in the chalk. The result was that the machine began to revolve instead, threshing around in thin air until it could be turned off. After boring a four metre diametre tunnel for 140 metres, the machine was abandoned in the cliffs near Folkestone. It is still there rusting and provided a nest for bees in 1984. In 1986 the Science Museum was trying to raise finance to salvage the machine, and some excavation has occurred.

In 1924, with the election of a minority Labour Government, the tunnellers tried again. Ramsay MacDonald referred the issue to the Committee of Imperial Defence, on which four previous premiers sat — Asquith, Baldwin, Balfour and Lloyd George. The military objection had been losing ground. Winston Churchill later said that even before the First World War the army chief of staff (Sir John French) and the first lord of the Admiralty (Prince Louis of Battenburg) both supported the tunnel. Churchill condemned the decision. "There is no doubt about their promptitude. The question is: Was their decision right or wrong? I do not hesitate to say that it was wrong."

In 1929 the tunnellers had another go, with a scheme that was a modified version of the 1906 plan — without the viaduct — and with the service tunnel running between the two main tunnels. The British Parliament set up a committee to consider the proposals.

There was the usual rush of lunacy. The year before, in 1928, William Collard, an Englishman, published his plans for a railway from London to Paris, running on seven feet gauge rails (the standard gauge in Britain and France is 4 feet 8.5 inches). He estimated the entire scheme to cost £200 million. Such was the grandness of the conception that the Channel Tunnel was but a small fraction of the cost.

By comparison the bridge, at £75 million, seemed a bargain. It carried a four lane road, and a twin track railway — together with a station in the middle, presumably for the fishing trade. An Italian engineer believed that a submerged tube, of iron encased in concrete would do the trick, for £2 million. But if the committee awarded biscuits, then the whole barrel would have been taken by the Swiss engineer, who wanted to build two causeways across the Channel (for 80 million) to establish a cross-Channel canal.

The committee, not surprisingly given some of the other options, supported the Channel Tunnel Company's scheme. The committee concluded that it "would be of economic advantage to the country." It ruled out any public sector finance. The Committee of Imperial Defence took its usual stand. It warned against "increased military commitments involving an element of danger, to provide against which a heavy capital and annual expenditure could be incurred."

Maybe it was hoping to turn the xenophobic vote, but the Channel Tunnel Company had a distinctly unusual way of justifying its own scheme. It said that 1.9 million passengers would use the tunnel, plus a further 1.7 million "foreigners".

In the House of Commons the Channel Tunnel Company's Bill failed by just seven votes. The vote produced some curious alliances. Both Clement Atlee and Herbert Morrison voted against. While Aneurin Bevin and Winston Churchill were amongst the 172 MPs supporting it.

The vote virtually killed the Channel Tunnel for the rest of the decade. Eventually in 1936 one of the schemes for improving the links between British and French railways was acted on, Sir John Fowler's plan for a train ferry. Fowler had proposed evening out the difference in height between the rails and the ship, because of tides, by hydraulic jacks. The Southern Railway adopted a similar solution, opening a tidal lock. Its two crack expresses, the Golden Arrow and the Night Ferry, took these train ferries.

In 1939, with war looming, the French again tried to interest the British in a tunnel. Neville Chamberlain, the Prime Minister, and son of Joseph, turned the approach down. With the invitation of France the British Government became worried that the Germans might build their own tunnel, and a special watch was kept on the site at Sangatte. The

The 1930s Whitaker tunnelling machine excavated a trial Chunnel boring in the chalk cliffs at Folkestone and remains there to this day. It was totally buried in a landslide at one stage but is now partially exposed, but the four-metre cutting head is still firmly embedded in rock.

engineer, Sir William Halcrow, advised the Government that the Germans would not be able to hide the spoil from the tunnel without anyone noticing. Later in the war the British thought of building a tunnel to help their invasion. The idea was droppped after the Government found that a tunnel would take eight years to build.

Interest in a tunnel revived immediately after the war. In 1947 a Frenchman, Andre Basdevant, produced an implausible idea for a bored tunnel. The single bore would carry four lanes of road traffic and two railway lines. The next year a Channel Tunnel Study Group was set up in British Parliament. Its secretary was Lt Commander Christopher Powell, who was also secretary of the Parliamentary Scientific Committee, and one of the most respected lobbyists in Parliament. The committee helped to pave the way for the 1960s plans. Meanwhile the surface crossing to the Continent became if anything slightly more difficult. In the 1960s hovercraft (and jetfoils ten years later) helped to cut an hour or more off the crossing. But in 1972 British Rail withdrew the Golden Arrow. The Night Ferry to Brussels and Paris lingered for another five years, until

1977. The journeys by train and boat from London to Paris now takes longer than it did before the First World War.

This doubtless provided the spur for England's engineering ingenuity. Milton Boothroyd wanted to build a £1.4 billion suspension bridge. The bridge has two decks. The upper deck has a six lane motorway, with space for three railway tracks. The lower deck is for trolleybuses, cycle lanes and a service area. The bridge would be assembled by airships, each capable of carrying 400 tonnes [39].

For £5 billion Eurolink, a group of businesses in London's east end, offer a double deck bridge for road and rail, with hydro-electric turbines generating electricity in the bridge piers and an international conference centre in mid Channel — to take advantage of the duty-frees.

Martin McCullogh's Leviathan is a plan for two causeways, to provide reservoirs for generating hydroelectricity. The causeways would be linked to each other, and to land by bridges so as not to impede navigation. The cost? £10 billion.

Eric Osmund Munday has the distinction of producing the most expensive plan for crossing the Channel. His steel railway viaduct would begin in Edinburgh, and run to Dover, via London. The railway would then dive under the Channel and continue on to places such as Paris and Lyon.

The trains would initially have a top speed of 270 kilometres per hour, the same as the Train à Grande Vitesse, in France, but eventually they would run at 500 kilometres per hour. The viaducts would be glazed and weatherproof. The lines would be straight so that Munday says there would be "no spilling of drink".

The one common factor to all these last four schemes is that they date from 1985. They were among the ten schemes submitted to the British and French Governments. Sadly they were never seriously considered. None put up the two £175,000 bonds demanded by the Governments - although Boothroyd did offer the the Grand Hotel at Folkestone, as security. It was not accepted. They also ran.

However millions of travellers every year have cause to be grateful that the eventual winner was not a drive through scheme. In the invitation to promoters the British and French Governments said: "For safety reasons it would appear desirable for the transistion from the general traffic rules applicable in France to those applicable in Great Britain, or vice versa, to take place as the user leaves the link…" [40] So the French would drive on the right and the British on the left.

References

[1]   *Sunday Times* and *Observer*, 19 January 1975.
[2]   The *Guardian*, 21 January 1975.
[3]   *The Times*, 21 January 1975.
[4]   Humphrey Slater and Correlli Barrett, *The Channel Tunnel*, Alan Wingate, 1958, p. 2: Thomas Whiteside has a slightly different account in the *Observer Magazine*, 8 October 1972.
[5]   *ibid*.
[6]   Channel Tunnel Study Group, *The Channel Tunnel*. The facts, Channel Tunnel Study Group, 1964.
[7]   *Prospects for a Fixed Channel Link*, Cmnd 2137, HMSO, 1963.
[8]   *ibid*.
[9]   *Hansard*, House of Commons, 6 February 1964.
[10] *New Scientist*, 11 October 1973.
[11] Major Projects Association, seminar on the Channel Tunnel the lessons of 1970 – 1975, December 1981.
[12] *The Channel Tunnel Project*, HMSO, Cmnd 5256, March 1973.
[13] *The Channel Tunnel*, HMSO, Cmnd 5430, September 1973.
[14] *New Scientist, op cit*.
[15] *New Scientist*, 1 November 1973.
[16] Major Projects Association, *op cit*.
[17] *ibid*.
[18] *Hansard*, House of Commons, 25 October 1973.
[19] *Sunday Times*, 19 January 1975.
[20] reported in *New Scientist*, 21 November 1974.
[21] Cairncross report, quoted in Major Projects Association, op cit.
[22] Crosland in conversation after addressing Transport and Environment Group meeting in 1975.
[23] The Ministry of Transport became part of the Department of the Environment in 1970; it was separated (as the Department of Transport) in 1976.
[24] Major Projects Association, *op cit*.
[25] *ibid*, the speaker was Peter Keen.
[26] Cairncross, *op cit*.
[27] Major Projects Association, *op cit*.
[28] *ibid*.
[29] British Rail press release, 20 September 1978.
[30] Department of Transport press release, 23 January 1981.
[31] Nord-Pas de Calais, La Liaison-Fixe Transmanche, Hier, Aujourd'hui, Demain, Lille 1986.

[32] *ibid*.
[33] Gosta Sandstrom, *The History of Tunnelling*, Barrie Books 1963.
[34] Besides the Nord-Pas de Calais, Sandstrom and Whiteside, all *op cit*, the history from 1830 to the 1950s is also covered in:
Deryck Abel, *Channel Underground*, Pall Mall, 1961.
Channel Tunnel Project, *The Channel Tunnel*, 1948.
Gavin Gibbons, *Trains under the Channel*, Advertiser Press, Huddersfield, 1970.
Peter Haining, *Eurotunnel*, New English Library, 1972.
A S Travis, *Channel Tunnel 1802–1967*, Channel Tunnel Association.
Thomas Whiteside, *The Tunnel under the Channel*, Hart-Davis, 1961.
Bouygues, *Petite histoire du tunnel*, 1987.
Reginald Ryves, *The Channel Tunnel*, Batsford, 1929.
[35] This letter is reproduced in Haining, *op cit*.
[36] *Nature*, 21 April 1870.
[37] *Nature*, 9 December 1869.
[38] *Illustrated London News*, 4 March 1882.
[39] *New Scientist*, 7 November 1985.
[40] Department of Transport, *Invitation to Promoters*, 1985.

Links into the future

Peter De Ionno

Nature never meant Europe to be traversed easily. Its Alpine heartland and the inlets that eat into its shores creating a "peninsula of peninsulas" have divided peoples and obstructed free movement as much as its history of fierce, assertive nationalism. The most dramatic benefit of the Channel Tunnel may be to turn Continental enthusiasm for unity into reality, drawing the farthest flung reaches of the Community together into a network of high speed roads, bold bridges and tunnels.

Almost every obstacle has spawned a band of engineers believing they have the answer of a crossing that will satisfy commuters and bring businessmen, tourists and prosperity in an endless stream as surely as money makes money. Now technology has caught up with and over-taken the natural challenges. The only limitation to links that have been, for decades, little more than good intentions, are the financial and political barriers.

The difficulties of cross-border traffic in western Europe are negligible when compared with the hard line drawn between East and West, yet every trans-continental truck driver has a repertoire of woe and delay caused by EEC customs obstructions, mounds of paper-work and strikes.

Estimates of the effects of psychological barriers posed by borders, as opposed to the visible barriers presented by terrain, averages between four and five to one, when the flows between points within countries are compared to the same distance involving border transit. It is a significant figure, but although improved transport links can generate valuable new business, it is by no means guaranteed that faster surface transport services alone will ever demolish national barriers completely[1].

The inevitable trend is towards a cats-cradle of high speed links that, by the first 20 years of the 21st Century, could allow a continuous road journey from Oslo to Tangier and beyond. The Channel Tunnel will make it possible to drive from Inverness to Istanbul and on through Asia without ever having to wait for a ferry or cope with seasickness.

The Channel Tunnel is the closest scheme to fruition, but the Scandinavian countries continue to toy with proposals for eventual road and rail connection between Denmark, Norway and Sweden with western Europe. The plan, called Scanlink, is technically feasible and there is a powerful lobby that says not only is it financially viable but it is essential to the region's continuing prosperity.

Plans to either bridge or tunnel under the Strait of Gibraltar continue to tantalise the Spanish and the Moroccans. A fixed link between Africa and western Europe has the blessing of the United Nations. Much basic research remains to be done but the Moroccans would like to see work start by 1995.

To the east, the Strait of Messina begs for a crossing that would end Sicily's isolation from the rest of Italy and help reverse decades of decline. Turkey has made its commitment to a third bridge across the Bosporus almost as a down payment on acceptance into the European Economic Community (EEC), though much of the traffic apparently generated by each new link is domestic. Farther afield, in Hong Kong and the USA, smaller scale tunnel technology and ambitious bridges have proved their worth as effective investments in essential infrastructure. They are catalysts to development and expansion of services.

Yet one of the world's longest tunnels, Japan's famed 53.8-kilometre-long Seikan has proved to be a "good idea whose time has passed". After 21 years of construction and escalating costs its market tired of waiting and now flies over it on journeys dramatically shortened by by the advent of cheap jet travel. The failure of the contractors to work to schedule and budget stands as a warning to any others who venture into the realms of vast construction. But the experience has not deterred the Japanese who are pressing ahead with an imaginative, but expensive, project to cut across Tokyo Bay with a bridge–tunnel combination, made essential by the choking pressure of the city's uncontrollable traffic. Another huge scheme that will ultimately claim nine of the world's 20 longest suspension bridges is underway as the Japanese Government, eager to extend the benefits of its burgeoning high technology industry to underdeveloped areas, builds a web of road and rail bridges across the Inland Sea between the main island of Honshu and neighbouring Shikoku.

In the distant future there are even more dramatic plans. The Japanese have never forgotten their intention of tunnelling under the Sea of Japan to create a fixed link with mainland Asia. The Japanese Imperial Army once visualised a vast railway loop that started in Siberia, took in Tokyo and arched south to end in Korea. It is a plan to stretch the imagination and transform the region's economies into an interlinked powerhouse. Historical and racial rivalries still conspire to ensure it never happens, but both countries are engaged in a ten-year programme of talks on the huge link.

Ideological differences and isolation stand in the way of a fixed link between the United States and the Soviet Union across the 79 kilometres of the shallow Bering Strait. Modern technology has made a tunnel possible, but whether such a massive engineering venture requiring close cooperation between the superpowers to create high grade communications to its portals, just below the Arctic Circle, could ever be achieved depends on the fragile progress of reconciliation between opposing political systems. If the area's vast untapped reserves of natural resources are required in the future then the economics of mutual self-interest could justify the enormous costs that would be involved. It would be the link to literally bring the world together.

Another 100 years will assuredly produce a world which people now alive cannot imagine and certainly none would recognise. Even a futurologist with the prescience of Jules Verne would be hard-pressed to forecast where the accelerating pace of technological, political and social change will lead. But there no is doubt at all that fixed links joining countries and even continents will be a fact of life. They will change the world, completing the communications revolution started last century with the first arrival of motor transport and air travel.

Linking Europe

One of the most powerful and persuasive voices calling for a comprehensive network of fixed transport links to bring all of Europe closer together comes from The Roundtable of European Industrialists. Their 1984 report Missing Links: Upgrading Europe's Transborder Ground Transport Infrastructure opens with the words: "On a continental scale, deficiencies in Europe's ground transport system constitute an effective barrier to European economic and social progress."

Laying immersed tubes on the seabed from specially constructed barges. This method was used in Hong Kong, was suggested for the Channel fixed link and will probably be incorporated in the Scanlink scheme.

"Europe can only become a thriving unified market if there is the freest possible movement of goods, people, capital and ideas." With visionary statements that look beyond the here and now of corporate or state balance sheets, the Roundtable has taken the high ground of economic debate, and united lobbyists, promoting single infrastructure schemes under an umbrella of business acumen and common sense. "The talk, however, has not been matched by action," said the Roundtable. "This is hardly surprising. Bridges and tunnels across mountains and seas are expensive, politically controversial and environmentally suspect. At the end of a decade of high inflationary growth, it would be remarkable if western European governments possessed the self-confidence necessary to take up such major projects, even if they were ideologically pre-disposed towards major public spending initiatives in the first place, which some are not. But even the opponents of schemes like the Channel

Tunnel and the Fehmarn Belt crossing seldom argue that such additional links between countries are undesirable, merely that they are too expensive, too risky or too politically problematical."

Their rueful dismissal of nationalistic western European politicians is a sad reflection on the lingering aftermath of the Seventies' years when booms turned to bust and rust in a decade of recession and fiscal crisis. Post-war attitudes of adolescent confidence have matured into gloomy pragmatism. Hemmed in by shortsightedness and haunted by the failures of grand schemes like Concorde and the 1974 Channel Tunnel attempt, British politicians, in particular, have remained incapable of imagination or frightened of arousing ambition. So instead of leading from the front, governments assume a guardian's role, so wary of being held responsible for failures they deter many from trying at all. The variety of attitudes to distinctly similar problems is one of the most intriguing aspects of any examination of the approaches to Europe's missing links. It is notable that the governments showing the most enthusiasm are the Moroccans, the Turkish and the Italians, administrations that see fixed transport links as immediate solutions to the problems of areas beset by grinding poverty. Where the financial benefits effectively amount to the icing on a generally prosperous cake, as with the Channel Tunnel or the Scandinavian links, the governments are more circumspect, almost indifferent.

A major project can only be considered if it is self-financing. Ultimately this becomes a sole criterion and questions of infrastructural benefit and work created are dismissed, just as the rallying calls of politicians retired to play the elder statesman are rejected for their source rather than their content. It is an arrogance that invites ultimate disaster. "From a strictly industrial point of view the creation of new fixed transport links between European countries would speed distribution, permitting more efficient manufacturing systems. This in turn would raise the capital productivity of European industry," said the Roundtable report.

But strengthening Europe's transport connections would do more than this. "It would be a force for integration, fostering more cohesive markets for all of business, not just those companies heavily dependent upon physical goods distribution," the report added. "Europe is a market of 360 million people still waiting, in some respects, to be served effectively by the continent's own industrial companies. In the failure to integrate markets, the fragmentation goes to the heart of Europe's most pressing economic and employment problems."

Great Britain will cease being an island the day the Channel Tunnel opens. The notion of a Channel Tunnel, a permanent all-weather, physical bond between Britain and the Continent, satisfies everybody.

Britain and Europe

For the British it has been the perfect excuse for an argument, for calls for inquiries, for forming committees and subcommittees and for resisting change at any price. For while its supporters can produce complex cost benefit analysis the opposition, not just the ferry operators who see a lucrative business being undermined, fall back on cherished arguments of independence and protection from hostile foreign interests. Best of all it gives excuse for the British to demonstrate their distaste and disdain for Europe, especially the French. For many the hesitation has been deeply subconscious; there has been a dull but reassuring consistency in remaining Little Englanders. Although modern weaponry has destroyed the oldest arguments that a Tunnel would be an invitation to an invader there is still psychological security in the notion that Britain will always be safe as long as its enemies are at arm's length able to attack only by air or sea.

Lord Pennock, chairman of Eurotunnel until early 1987, recognised that the antipathy to the Channel Tunnel ran deeper than just commercial interest: "Our opponents have exploited something in the British mind which tells us we are an island race. Deep down because of people like Napoleon and Hitler there has been a feeling that we should not be joined to Europe. The European Economic Community has ensured that we will never again go to war against the nations of Europe, I am convinced of this."

Of course there is no threat of invasion, other than from language students and American shoppers cashing in on exchange rate fluctuations. If the Tunnel opens vistas beyond Britain's shores and accelerates the osmosis of cosmopolitan tastes and perceptions into the nation's traditionally xenophobic consciousness it will have been worth every penny.

It has been the British tendency to assume superior postures that alienate the nation from the Europeans, who have had to learn each other's languages and customs and adjust to each others needs while island Britain maintained aloof, distant.

1973 was a momentous year for Britain and Europe. It was the year Britain was admitted to the European Economic Community. It was also the year Europe was hit by the oil crisis. The year when the nations of Europe formed their agreement to create a powerful European political union by the end of the Seventies, while they tried to ensure their own survival and stop the slide into recession. Hardly an auspicious start for Britain and her new partners. Since then Britain has tried to be a good, if

Artist's impression of a bridge across the Strait of Messina. It would have the longest span in the world but with a deep catenary, might not be able to carry rail traffic. Either the bridge or its principal rival, a floating tunnel, would put Italian civil engineering ahead of the rest of the world . . . if construction ever starts. Source: Stretto di Messina SpA.

somewhat unwilling European. The Conservatives buoyed by industry's overwhelming endorsement of the move into Europe paid increasingly enthusiastic lip service. "The EEC must be Britain's future," pronounced Mr Francis Pym, UK Foreign Secretary, on the tenth anniversary of admission while fending off the carping of the Labour Party about wine lakes and butter mountains[2].

Though the Thatcherite Conservative approach has been predominantly one of value for money, there is still room for loftier aspirations from the British in Brussels. Sir Henry Plumb, chairman of the Conservative Democratic Group said the difficulties Britain has been alleged to have and to foster in the forums of Europe were often exaggerated: "The clang and din from Brussels is not really the sound of battle, it is the noise and dust you will find on any construction site" [3].

Even the dull but apparently thoughtful Sir Geoffrey Howe saw a role for Britain in helping the Community grow in stature when he spoke to Scottish Conservatives: "We want to make the Community not just a force for stability, democracy and prosperity in Europe but for liberty, peace and civilised relations between nations around the world. We want greater unity in Europe to help strengthen western security and to open up new avenues of contact between east and west. We want Europe to build on its unique links with the Third World, to increase understanding and cooperation between developed and developing nations. We want to see the Community use its vast resources of technical know how and skilled manpower to reassert its position as a powerhouse of technological and commercial success" [4].

No one doubts a new shape for Europe is being forged in the European Economic Community. Neither is there any doubt that the development of Continental Europe will have a remembrance for the past running alongside a passion for the future, a mixed sentiment that puts it in conflict with the British ossified in island isolation. It has been no surprise that, despite increasing cooperation at the political level, the popular perception that the EEC is an alien association of old enemies and rivals whose combined interests are generally at odds with Britain's has been almost impossible to lay to rest when the nation's most popular newspaper the *Sun* selling more than 4 million copies a day, boasts its disreputable xenophobia across the front page. It is a shallow view ignoring the weight of history.

Britain has no other place in the world besides Europe. Europe has been a mainspring of intellectual, material and technical development for more than 2000 years. It has often produced tyranny as an answer to crisis; there are tears in the economic fabric and there are injustices but the underlying movement has been towards civilisation and away from barbarism. It is European, not just British, tradition that has nurtured and matured political thought and produced the moral milestones of justice and emancipation. The abolition of slavery and the evolution of democracy have enshrined freedom as a code by which to shape the world. The British apart from having shared ancestors, experience and culture with Continental Europeans suffer the same fate of being locked into the Old World between the crushing power of the United States and the Soviet Union.

Likewise thrusting nations sprung from ambitious empires established by mercantile western Europe have become the Third and Fourth Worlds, rearing up with industrial might and commercial ambition against Europeans comfortable with complacency.

Now Western and northern Europe, split from the East in the tense aftermath of World War II, is effectively a gathering of communities cleaving together for reassurance. Alone, they are utterly vulnerable to the superpowers, who offer alternatives of bribes for their clients and military intimidation for their enemies. And there is a direct threat from the expansionist new economies of Japan, Korea and Taiwan in the Far East. In the coming years, Brazil, India and China, thus far under-developed giants, will show powerful industrial competition to the rest of the world as the move to exploit their low wage economies accelerates from the Pacific Basin as well as Europe and the United States.

Railways: yesterday's idea with tomorrow's technology

One fact is certain. Trains will dominate the high-speed transport corridors of Europe into the 21st century and beyond. Oil price crises, high costs, relatively high accident rates, overcrowding and environmental concerns about emissions all combine to suggest a future where the motor vehicle will be less viable and less welcome than it is today. The citizens of the next century are likely to find escape from the stress and pressures of personal motoring as the train takes the strain.

Technology has been helping the railways of Europe make up the passenger and freight business they have lost to the motor vehicle and to air travel in the past three decades. Direct rail links with express trains capable of 350 and 400 kilometres an hour are not far away and by 2020 they may even be left standing by high energy, environmentally pure MagLev (magnetic levitation) and induction loop trains capable of doubling and trebling those speeds. One imaginative proposal for a direct underground link between London and Paris estimates that door to door journeys can be cut to 30 minutes. The question of establishing advanced rail services to criss-cross Europe has been treated increasingly seriously as higher speed operation has become a reality.

The success of the French Train à Grand Vitesse has invigorated interest in passenger rail travel with the cutting of the time for the 425 kilometre route from Paris to Lyon from four hours to two hours. It runs at a profit, unlike the Japanese Shinkansen, the Bullet Train, that revolutionised passenger traffic patterns in Japan but still runs at a loss. British plans for the Advanced Passenger trains failed because of protracted engineering setbacks to the system that tilted the carriages on bends. But the Germans are enthusiastically testing magnetic levitation on an experimental monorail, the TR06 vehicle which has already achieved speeds of 300 kilometres an hour and is intended to operate at 400 kilometres an hour. The Italians are busy working on the Direttissima which is to run between Rome and Florence at 250 kilometres an hour. The enthusiasm for the TGV and realisation of its potential for the Continent as a whole was a factor in the French lobbying for the Channel Tunnel. After more than a century of rail transport the operators are looking over the back fences of their borders and finding ways to meld and unify four different traction systems, a multitude of signalling and communications systems and a range of track widths. All they need to make it happen are rails across, through and under the obstacles. The rest is easy.

Scandinavia

It takes a car about 20 hours to travel between Oslo and Hamburg by a combination of highways of varying quality and two weather-dependent ferry crossings. To cover the same 725 kilometres on mainland Europe's motorways takes less than ten hours. This stark comparison is at the heart of the transport development plan, dubbed Scanlink, a blueprint to integrate Scandinavia with Europe along an unbroken transcontinental corridor of high-speed motorways and railways. Sitting high on the northern fringes of Europe, the Scandinavian countries are dependent on more populous southern countries for trade. But as the mainland motorway network has grown and high-speed rail services have been developed, the Scandinavian countries have been left farther out on their geographical limb separated from the economic hurly burly of western Europe by tantalisingly narrow sea channels.

But even though they sit at the crossroads between northern Scandinavia and Europe, there are a great number of Danes who do not want fixed links from their country of many islands to anywhere. They have grown up with ferries as part of their lives. The enforced delays and

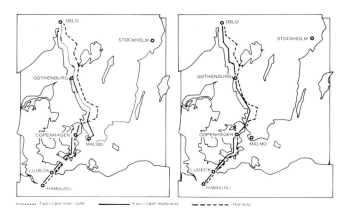

Travelling from Oslo in Norway to Hamburg in West Germany, without (left) or with (right) the Scanlink scheme.

inactivity waiting and sailing between road connections agrees with their temperament. However, in the Danish Parliament, the Folketing, with all parties agreed on the need for a drive-through link with Europe, more than three-quarters of the members have been in more of a hurry. A third and final reading of the Bill authorising construction of a link across the 60-metre deep Great Belt waterway between Copenhagen and the Danish mainland, the Jutland Peninsula, was achieved before the summer recess in June 1987.

The Danish government's decision [5], announced in June 1986 for a $1.56 billion drive-through bridge and tunnel link across the 18-kilometre Great Belt, the entrance to the Baltic Sea, between the islands of Zealand and Funen, may have scuppered the grander plan of the Scanlink vision. This put the connection to the rest of Europe across the Fehmarn Belt to the south, creating direct access to northern Germany and included a link from Copenhagen to Sweden, the Oresund connection. Supporters of the Scanlink idea argue that it would end Scandinavia's isolation at a stroke, starting a process to improve communications throughout the continent that would quickly become impossible to stop[6]. But the go-ahead for the Great Belt means the dream of links to enable road vehicles to drive unimpeded from Sweden to Europe may have to wait a decade or more.

There may still be some hope for a plan to link Denmark and Sweden with bridges and tunnels, but the Great Belt scheme does seem to have destroyed any hope of a bridge link between Denmark's island of Lolland and West Germany, dubbed the Fehmarn Belt scheme in the Scanlink package. The Danes believe that it would compete for business with the Great Belt connection, leaving both links with too little traffic to service the vast capital borrowings involved. So the Fehmarn Belt project lies dormant and shows little sign of revival.

Plans to bridge the 16.6 kilometres of water have been under consideration since 1962 when the Fehmarn–Lolland group, a gathering of bankers,
industrialists, chambers of commerce and the West German Transport
Ministry suggested a four-lane road bridge and a dual-track railways. A
tunnel solution was also examined and pronounced realistic. Cost for the
project which would face similar conditions and water depths to the Great
Belt proposal was estimated at $1.5 billion. The Roundtable of European
Industrialists examined the package of plans in 1984, concluding that the
total cost of the Oresund link and the Fehmarn link, together with
essential road and rail upgrading for a four-lane motorway and dual-track
high-speed railway between Oslo and Hamburg, would be $4.15 billion.
A lesser option of only building motorways where they have already been
planned and not improving the railways would be $2.7 billion.

The Roundtable said: "The case for various links in Scanlink is
enhanced by examining transport patterns across the entire geographical
region. As a regional system Scanlink has higher traffic potential than the
sum of individual projections which have been made for its various fixed
links." Appreciative of opposition from those daunted by the cost they
suggest it should be a staged project. There is more than an element of
cunning in their suggestion: "The key to Scanlink initially is to build the
sea crossings. Once this is achieved Scanlink will in a sense become
operational. The build-up of traffic caused by the completion of the fixed
links, however, would create the momentum needed for the upgrading of
related road and rail systems." So far the commitment to the Great Belt
idea has done no more than spark a new round of talks between the
Swedish and Danish governments about a bridge, embankment and
tunnel scheme to join the two countries.

The link across the Baltic Strait, the 0resund, was first investigated by a
Swedish–Danish Committee in 1978 when routes and design standards
were considered. The preferred option was a four-lane motorway
between Copenhagen and Malmo on the Swedish side. Passenger trains
would follow the same line hopping from the Danish mainland to the
islands of Amager and Saltholm. Rail freight would cross between
Helsingor and Helsingbord. The joint commission reported in 1985. The
latest working group is due to report in July 1987. Lobbying for the the
project is mainly Swedish but their Norwegain neighbours have also
offered financial assistance. In Denmark enthusiasm for a second vast
enterprise is muted. The Great Belt project has come under constant
attack for being an unnecessary expense.

"The link between Denmark and Sweden is entirely for the politi-
cians." said Soeren Tengvad, head of planning for the Danish National
Road administration. "The Danish people would only be interested if
Sweden was to fund the greatest part of it. As we see it the main traffic is

Swedish. The road to Denmark leads to Europe, Sweden is only the way to Denmark." There is every reason to believe that once the Great Belt link proves itself, the lure of connection will prove irresitible to the Swedes. They will not be able to ignore the benefits. Oresund alone would create 12,500 man years of work equivalent to creation of 2000 direct jobs.

But the Great Belt Project itself is ambitious. It will create 25,000 man years of employment. Though when it is completed there will only be 400 jobs to take the place of the 1800 people now employed on the ferries that will all but disappear by the mid 1990s. The Great Belt is Denmark's Channel Tunnel, complete with a history of proposals and failed attempts since 1973. But there has never been any doubt it would transform travel and distribution economics within Denmark, cutting times between Copenhagen, the capital, and other cities by 30 per cent. The 300-kilometre trip from Copenhagen to the second city of Aarhus now takes five hours. After the Great Belt motorway it will be around three and half hours. Construction of the first phase of the railway link which includes a ten kilometre tunnel from Zealand to the island of Sprogoe will start in 1988 and is expected to be operating in 1993. The motorway will also include a bridge across the busy sea lanes between the Baltic and North Seas with a span of 1500 metres or two spans of 800 metres. The bridge will be 14 metres above sea level rising in an arch to 80 metres, to give clearance for the largest vessels. Estimates of motorway traffic suggest that up to 40,000 vehicles a day will use the link by the turn of the century. The Danes have scorned a Eurotunnel-style rail piggyback system for vehicles in favour of separate rail and road solutions in the face of intense pressure fron the motoring lobby. The route will add 160 kilometres to the journey from Copenhagen to Hamburg and beyond, but drivers will still take about the same time for the trip.

Every variant of crossing has been examined: bridges, bored tunnel, submerged tunnel and causeways. The decision on whether the tunnel will be bored or an immersed tube will be taken by the contractors in their tendering. The scheme will be privately funded by a combination of institutional investment and by public flotation.

The Great Belt is the only entrance for shipping to the Baltic Sea ports of Russia, Poland and East Germany, apart from the Kiel Canal. Although the shipping traffic is heavy the link supporters, with the enthusiasm all visionary engineers share in abundance, maintain that the reduction in ferry services crossing the main sea lanes would bring a dramatic improvement in safety. The car ferries will continue to operate between 1993 and 1996. They contend too, that the web of bridges and tunnel ventilation shafts could be constructed in ways that would make them only minor hazards and obstacles.

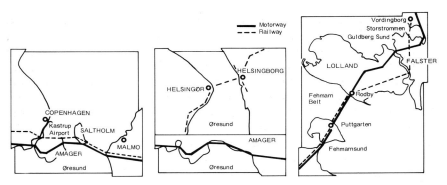

Different options for crossing Denmark in the Scanlink scheme.

Scanlink is optimistic too, on the traffic forecasts, the crucial factor determining whether the crossings will ever be built. Up until 1984 traffic estimates had been understated because they failed to incorporate the benefits of links attracting new traffic and diversions from more distant ferry services, the Roundtable said. They estimated that up to 90 per cent of Sweden's rail traffic between Western Europe would be routed through Denmark rather than the half that now trecks across East Germany. Using a macroeconomic formula of real gross national product growth of 2 per cent a year, real disposable income growth of 1 per cent, freight traffic growth at 2.5 per cent and passenger increases of 1 per cent, the Roundtable produced estimates for the year 2000 (see table 1).

Assuming that these figures, averaged for 2000 on the basis that they would initially be overstated, and in later years too low, the Roundtable concludes that the annual revenues, on charges similar to those of the ferries, would be $129.1 million, $132.6 million and $205.4 million respectively. Some 80 per cent of income would be raised from commercial traffic. Estimates of annual operating costs for each link would be only $3 to $3.5 million. On these figures the projected rate of return on capital is around 13 per cent. The Roundtable, consistent with its general commitment to development independent of State funding, suggested finance could be raised in a similar way to that for Eurotunnel, with loan and institutional capital to launch then a public share offer to follow.

Gibraltar

As with all international link schemes the 28-kilometre crossing of the Strait of Gibraltar between Spain and Morocco could be built with the paper generated by the reports and feasibility studies and supported by the hot air from the politicians talking about it.

But despite the talk, there have been no decisions. The plan would please the EEC, particularly with the new membership of Spain, as it would create a physical connection with continental Africa, bringing it within the Community's economic orbit. The Moroccans too are keen to cement their ties with Europe. Like the Channel, the Strait is among the busiest shipping lanes in the world and has been a natural crossing point between Africa and Western Europe for centuries. And again, as with the Channel the challenge of bridging it has been a lure to engineers for decades. The most conservative estimates of traffic predict 13 million people crossing the Strait annually by the year 2000.

The likeliest place for a crossing has been identified as from a point east of Tangier to Tarifa, west of Gibraltar [8]. At the end of 1986 construction costs were estimated at $3 billion [9], with the project taking between 10 and 14 years to complete. The idea of a tunnel or bridge across the Strait was first raised in 1979 by the Spanish but it has been most enthusiastically pursued by the Moroccans through the Societe Nationale d'Etudes de Detroit. The United Nations has given approval in principle to the link but data on the geology of the crossing are sparse. Judging from the list of works, geophysical surveys including collecting of rock samples, a seismological programme, a hydrodynamics survey of currents and swells, surveys of the continental shelf, land surveys, traffic and economic impact surveys and the inevitable legal and financial investigations, it may be at least another decade before the decision-making proceeds further.

As the depth of the Strait is only 350 metres at maximum, a bored tunnel is feasible, but proposals have already been mooted for a tunnel, a suspension bridge and a combination of both. Again as with the Channel Tunnel the all-rail option is favoured on cost grounds over a drive-through tunnel. A road tunnel would cost about half as much again as a rail tunnel and present ventilation difficulties and take almost 40 per cent longer to build. The bridge proposal would present particular problems in piling the support tower foundations, due to the expected strength of the currents surging between the Mediterranean and the Atlantic. During World War II submarines of both the Allies and the Axis powers used the distinct but powerful current layers caused by salt saturation of the waters of the Mediterranean, where evaporation rates were higher than in the Atlantic to travel silently through the Strait to avoid detection by surface ships. The currents may also affect plans for a floating bridge anchored to the sea-bed in the way that North Sea oil rigs are fixed. It would leave clear lanes for shipping at either by descending into immersed tube tunnels. A floating tunnel has also been suggested.

If the talking and planning goes on as long as expected it may provide time for the advent of new materials such carbon fibre reinforced plastic

(CFRP), offering the same strength as conventional building materials with the bonus of an 80 per cent saving in weight. Swiss engineer Mr Urs Meier believes a bridge contructed almost entirely of CFRP would be possible within 30 years.

Because of the vast spans required, concrete and steel bridges could collapse under their own weight. Meier believes that 46,000 tonnes of CFRP decking supported by another 104,000 tonnes of suspension cables, each six millimetres in diameter and containing 500,000 carbon fibres, would be required to build a bridge carrying six lanes of road traffic and two railway tracks across a span of almost 8.5 kilometres. The two-deck structure would be held together by a lattice of CFRP tubes and beams creating immensely strong box sections. An additional benefit of using such high-technology materials apart from their high tensile strengths would be resistance to the corrosion that afflicts convential material structures in areas of high exposure to salt spray.

Messina

With the exception of the Channel, few waterways have figured so large in history and literature as the Strait of Messina. The three kilometre stretch of water has been the stuff of legend since ancient times. Homer made it the home of supernatural hazards that threatened to engulf Odysseus and other wayward mariners. Scylla was a rock on the Calabrian side and Charbydis was a whirlpool near Messina.

The first plans to bridge the narrow, turbulent sea passage between the toe of Italy and Sicily were made by Hannibal. He considered a pontoon bridge across the Strait, so that after he had transported his elephants from Carthage on the north African coast to Sicily, there was easy access to the Italian peninsula and straight to the back door of Rome, instead of across the Alps. Napoleon wanted to bridge the Strait and so did Garibaldi when he reunified Italy last century.

Modern plans to link the island with the European mainland and stretch direct communications closer towards north Africa began with a propc~al from Italian engineer Carlo Navone for a rail tunnel in 1870. The rapid economic development of northern and central Italy since World War II has only highlighted the economic blight and decline of Calabria and particularly Sicily cut off by a transport bottleneck at the Strait [10]. Despite attempts to streamline the ferry crossing the average time taken is about two hours for railway rolling stock and about one and a half hours for road vehicles, almost the time for travelling the 200-kilometre motorway between Rome and Naples.

Scanlink in the European context.

In 1985 more than three million road vehicles and railway trucks, together with six million tonnes of cargo, were ferried across the Strait. Passengers totalled 12 million and the traffic is increasing steadily. By the mid-1990s the passenger figure will have risen to almost 22 million and the numbers of vehicles to 5.25 million, according to the Stretto di Messina, a public company established by the Italian Government in 1971 to construct and operate a crossing.

Mr Claudio Signorile, the Minister of Transport in the Government of Mr Bettino Craxi summed up national commitment in January 1986: "We have a civil and economic obligation to build it" [11]. The European Parliament supports the project for the same reasons as it has backed the the Channel Tunnel as a means of alleviating and correcting regional imbalance.

But apart from the obvious commercial benefits of a fixed link with Sicily, removal of the ferries which currently cross the Strait 83,000 times a year would assist the promotion of the channel as a preferred maritime route between the southern French ports, the Tyrrhenian ports of Italy and the Ionian and Aegean Seas, the Bosporus and the Suez Canal. The Italians believe that with full integration of their northern inland waterways a dramatic circumnavigational route around the Italian peninsula could be established. Physical barriers faced by the engineers charged with building the link are immense. The passage is swept by violent winds that can reach 100 kilometres an hour, the Strait and hinterland are crossed by active seismic fault lines and the strong tidal currents reach a peak of six knots, equivalent in force to a 200-kilometre-an-hour wind. The geological problems are severe; sitting astride the intersection of the European and African geological plates, the area is one of the world's most earthquake prone regions. In 1908 an earthquake with its epicentre almost in the centre of the Strait levelled Messina killing 84,000 and started a six-metre high tidal wave that ravaged coastlines more than 200 kilometres away. Designs for a crossing will have to be able to withstand quakes registering up to nine on the open-ended Richter scale. In 1950 the Italian Government paid Steinman Boynton, Gronquist and Birdsall of New York to design a crossing [12]. They came up with a suspension bridge with one tower in the water on the Italian side. The design has since been modified to put both towers on land.

In 1969 an Italian Government sponsored international competition for new designs attracted 154 entries, of which six were shortlisted and subjected to feasibility studies, winning £10,000 prizes. Options included suspension bridges, a sunken tunnel and a British engineer's idea for a revolutionary floating tunnel.

In 1987 the competition had come down two designs, a giant suspension bridge, the world's biggest, designed by Italian engineer Mr Sergio Musmeci and the floating tunnel by Englishman, Mr Alan Grant. An Italian Government commission is considering feasibility studies for both schemes and a much delayed decision was expected in 1987, so detailed designs and preparations could begin within a year. The $3 billion plus road and rail suspension bridge would be breathtaking, with a 76-metre high main span of 3.2 kilometres, more than twice the UK Humber Bridge's world record central span of 1410 metres. On either shore the support pylons would rise to 381 metres, the height of the Empire State Building, in New York. It would take about eight years to build. The bridge plan has many supporters in the Italian Government but doubts linger over whether it will be able to carry rail traffic over such a catenary, a crucial factor, and whether a structure of such scale would be able to withstand buffeting from high winds, though the designers claim a model has been wind-tunnel tested to speeds of more than 240 kilometres an hour.

Alan Grant called his design the Archimedes Bridge because it relies on the principle of buoyancy discovered by the Greek mathematician who once lived in the city of Syracuse on Sicily. The tunnel, a concrete tube containing sections for both road and rail traffic, would float like a submarine about 30 metres below the surface and be held in place by a web of counterposed anchor cables fixed to the sea floor. The structure would be built in dry dock and towed to the site by tugs before being sunk into position. Because it has a natural buoyancy, loads inside the tube would ease the stresses on the cables enabling it to theoretically be used across distances previously considered unbridgeable. Grant claimed the structure would be virtually unaffected by earthquakes and resistant to the fierce tidal flow not bending more than a metre over the proposed length of 3.5 kilometres, whereas a suspension bridge would move up and down as much as six metres under load from traffic and much more in severe winds. The floating tunnel has the support of environmentalist groups because it would be invisible, preserving the area's scenic and historical integrity. Selection of the $4 billion Archimedes Bridge proposal could signal adoption of the principle for fixed crossings in many other parts of the world, including Gibraltar and the Gulf of Bothnia between Sweden and Finland. The design is being considered for a 2 kilometre crossing between Rion and Antirion at the narrows of the Gulf of Corinth in Greece, a 2.25 kilometre link in the USA between Plattsburgh and Burlington Vermont, across Lake Champlain, also in the US and for a 1.25 kilometre crossing of the Eidfjorden in Norway.

The Italians are committed to bridging the gap between Europe and Sicily, reaching out towards North Africa. All that remains is for political will to be translated into commercial and engineering commitment. In the words of Stretto di Messina the crossing will be a "testimony to the creativity and determination of mankind." But it just might be unseen.

Japan

The Japanese are justifiably renowned for their ingenuity. In using every available space on their overcrowded islands they have demonstrated a readiness to undertake complex engineering. It has enabled them to exploit and colonise limited and difficult terrain with enthusiasm that owes as much to necessity as it does to enterprise. Their tunnellers and bridge builders have to take into account a volcanic geology often subject to earthquakes and resulting tidal waves. On the surface, typhoons threaten fixed structures as well as sea traffic.

The Seikan Tunnel is the world's longest public transport tunnel, running 53.85 kilometres under the Tsugaru Strait between Honshu and Hokkaido. It is one of the engineering wonders of the world. Yet it is a failure. Only a year after tunnellers, working from opposite sides, linked 100 metres below ground, the Seikan was being branded the "idea whose time has come, and gone"[13]. Conceived in the age of steam and completed in March 1984, after 20 years and ten months of work, the Seikan is the model to prove that the Channel Tunnel is technically feasible, particularly as it will be cut through far more forgiving and manageable geology. The Seikan Tunnel is all that is left of a grandiose scheme once part of the Imperial Japanese Army plan for East Asia [14]. The generals wanted to attach Japan to the Asian mainland by stringing Honshu and Hokkaido onto a circular railway joining up the Soviet island of Sakhalin in the north, and Korea in the south with Tokyo on the eastern rim. A tunnel under the 85-kilometre wide, but relatively shallow, La Perouse Strait would be the main element in a link between Japan and Siberian Russia. In the south there has long been talk of a tunnel between Japan and Korea but the distances are daunting. It is 100 kilometres across

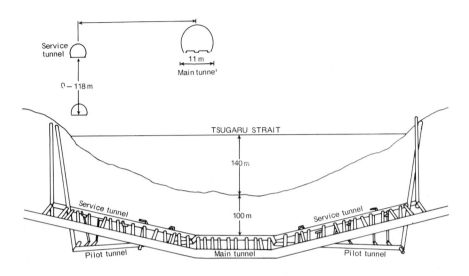

Cross-section of the troubled Seikan Tunnel in Japan, still not in use 20 years after
construction started. The difficult geology proved the Channel to be an easy option in
comparison.

the waters of the Nishi-suido from Pusan to Tsushima, a substantial
island straddling the Korea Strait, and further 85 kilometres from
Tsushima to Ik, another Japanese island. A third stretch of water between
Ik and Kyushu is 61 kilometres wide. The engineering scale and
complexity of the problem are so large and involved, it does not seem
likely that it will be seriously attempted for at least 30 years but the
Japanese Government still claims to be having talks on the subject with
the Koreans. If only one half of this circle was completed Japan's
dependence on the sea for access to her vital raw materials would vanish.
It is a dream that might never be realised but if it does happen it would
transform the region and perhaps deflect the pressure of Japanese
expansionist trade policies away from the West to the East just as the
Imperial generals wanted before World War II.

The push for the Seikan Tunnel started again in earnest on 26
September 1954 when a typhoon, roaring down the Tsugaru Strait, sank
five ferry boats including the Toya Maru, which gave its name to the
disaster that claimed 1430 lives.

Two years later, a Government study pronounced the idea of a tunnel
technically feasible, announced a contruction timetable of ten years and a
budget of 60 billion yen ($167 million at 1956 exchange rates). It took twice
as long as projected to build and if the dependence on rail travel was just
starting to fade in the mid-Sixties as airline travel and private car

ownership rose in response to rising affluence, it had gone by 1984. The Ministry of Transport had always estimated that the tunnel would be used by 13.5 million passengers a year and carry 8.5 million tonnes of cargo each way. But on completion air travel had whittled those figures down to 2.1 million passengers and less than 3.5 million tonnes of freight [15].

The tunnel was to replace the Seikan ferry which even in 1987 was expected to be out of service 80 days a year because of storms. In 1956 the journey was even more dependent on the weather and the crossing from Aomori on Honshu to Hakodate on Hokkaido took four and a half hours. The Seikan Tunnel would enable a single rail journey to slash that to one hour and 20 minutes or just 50 minutes by the famed Shinkansen, the bullet train. Train and ferry journeys between Tokyo and Sapporo the main city on Hokkaido still take almost 17 hours. The tunnel was intended to cut that journey to six hours. By the time the Seikan was ready after a series of delays its market had tired of waiting. While the tunnellers toiled to meet the changing specifications of their political leaders, the public had learnt that travelling between the islands by jet aircraft was convenient and reasonably cheap. Even the most favourable figures for the tunnel trains put the price of a ticket only about ten per cent below the cost of an airfare. The flight from Tokyo to Sapporo takes just an hour and twenty minutes [16].

So far it has been a financial disaster. It will be years before the tunnel fully proves its worth. By the time the tunnellers linked, and celebrated with the ceremonial smashing open of barrels of saki, the bill had reached $3.5 billion, more than ten times the original estimate. Heavy capital borrowings were being repaid at $400 million a year, including interest. Operating costs for trains due to run through in late 1987 were estimated at $45 million and some $750 million was needed to strengthen railway tracks and the road-bed. But it was the refusal to pay the additional high cost of upgrading the line between Tokyo and Sapporo to handle bullet trains that has destroyed the viability of the tunnel. For without the dramatic time savings the Shinkansen would bring there is little advantage for any except local travellers. The track was narrowed by 36.8 centimetres to the gauge used by local trains, a move taken by many as a final admission that the tunnel cannot be made to pay.

A potential major customer for the operators is the military. The Japanese Defence Forces have four divisions stationed on their northern extremities facing the Soviet forces on Sakhalin, and the tunnel offers a protected reliable supply line. However, they have a small hurdle to get over before the quartermasters' trains start running; Japanese law prohibits the transport of munitions through tunnels. The Seikan contractors were still putting the finishing touches to their achievement in

1987. The 23.3 kilometre undersea section is itself longer than Europe's longest tunnel, the Simplon, which is a mere 20 kilometres under the Swiss Alps between Switzerland and France. The engineers' challenge was not so much the length of the Seikan, though that is in itself remarkable, but the variety and difficulty of the geological conditions [17]. There were 34 deaths and four major cave-ins. At its peak the project employed 3000 workers.

There are three tunnels. A narrow exploratory bore that the contractors, the Japan Railway Construction Corporation, started in 1964 was a pilot tunnel used to determine the geological conditions ahead. Improving techniques enabled the tunnellers to increase the distance of their probe from 300 metres to 2150 metres, giving the excavators up to two years warning of problems. Now it has a dual role of ventilation and drainage, it is up to 118 metres deeper than the main tunnels. On the Honshu side alone, permanently operating pumps suck out 24 tonnes of earth and sea water a day.

The second tunnel is a service tunnel about a quarter of the diameter of the 11-metre main tunnel above. The Seikan Tunnel has been cut and blasted through a succession of difficult strata with faulted beds at each end, sedimentary to the north, soft rock and sandy mudstone at the centre and igneous andesite in the southern third. The work was most difficult at the centre where the soft mud was prone to collapse. A single 500 metre section took 4.5 years to tunnel through. The tunnellers had to cope with nine serious faults allowing water into their drives. The largest cave-in, in 1976, took 70 days to control after water poured into the tunnel at the rate of 70 tonnes a minute. A crucial technique developed by the Japanese contractors was grouting with mixtures of cement and sodium silicate, which was pumped under high pressure up to 70 metres ahead of the cutting face, forming a concrete cone to stabilise the rock. At times it was necessary to grout up to six times the radius of the tunnel, slowing progress to as little as 35 metres a month [18]–[19].

Despite the obvious financial difficulties, the Seikan Tunnel is now a symbol of intense pride to the Japanese. Mr Yoshinori Sekiguchi, an engineer who spent a decade of his life constructing the Seikan said: "Linking Honshu and Hokkaido is worth more than money — it is the dream of the Japanese people. In the future it will give our children confidence in the greatness of Japan"[20].

Tokyo Bay

Anyone who has experienced the tortuous delays that turn the huge Tokyo expressways into solid day-long traffic jams will appreciate the

The most likely future physical links in Europe, barring political problems, would be
across: the Strait of Gibraltar; from Sicily to Italy; from Rion to Antera in Greece; the
Scandinavian link from Denmark to Sweden; another link across the Bosporus and even
a link from Scotland to Northern Ireland.

benefits offered by a plan to build a combination of bridges and tunnels
across Tokyo Bay [21]. The imaginative $6 billion scheme has dramatic
similarities to the rejected EuroRoute proposal for a drive-through bridge
and tunnel crossing for the Channel. The backers, a consortium of private
construction firms, local government and the Japan Highway Public
Corporation, propose to cut travelling times from Kawasaki, about halfway between Tokyo and Yokohama on the north-western shore of the
bay and Kisarazu, in Chiba Prefecture, a sparsely settled developing
region close to Narita International Airport on the south-eastern shore,
from 90 minutes to only 15 minutes by car.

The Trans-Tokyo Bay Highway, a 14.5 kilometre four-lane crossing
with a capacity of 30,000 vehicles a day, would start with a ten-kilometre
tunnel starting on the Kawasaki side and rise to a 6.5 hectare man-made
island about five kilometres from the opposite shore. A second man-
made island would house vital ventilation equipment. The scheme has
been under discussion for 20 years but now with the prospect that it could
be a catalyst to new development that could create 25,000 jobs it is
considered economically justifiable. Early plans were for a bridge but
after a 1983 survey showed that the bay's six major ports served some 1300
ships a day it was considered prudent to opt for the combination of tunnel
and bridge. It is expected that the tunnels will be housed in two 13-metre
bores cut through soft sedimentary material just 14 metres below the sea
floor.

The Inland Sea

The Japanese economic miracle has been generally confined to the coastal
strips wrapped around the backbone of mountains on the main island of

Honshu. Shikoku, tucked into the scenic waters of the Inland Sea south of Honshu, is in the process of being firmly linked by road and rail to the rest of Japan [18]. The Government project is to help attract high technology industry to the predominantly agricultural island.

Three lines of bridges, two of them carrying road and rail traffic, are under construction in a vast project estimated to be costing $19 billion. Two bridges are open between Imabari in the north-west of Shikoku and Onimichi on Honshu. A third is due for completion in 1988. At the eastern end of Shikoku a suspension bridge from Naruto to Awaji island for the link to Honshu near Kobe has been finished. In the centre the Kojima to Sakaide route will carry bridges with central spans of 1100, 990 and 940 metres. When completed by the turn of the century the project will have created nine of the 20 longest suspension bridges in the world. As the Inland Sea and its 950 islands have been designated a national park since 1935, the choice of bridge type and the route were influenced by environmentalists who sought to keep to a minimum the visual impact of the works.

Turkey

The Bosporus, the narrow Strait between the Black Sea and the Sea of Marmara, not only divides Istanbul but it is the last water barrier on the road east from Europe into Asia. Turkey is one of the fastest developing countries in the Mediterranean region and it should be a full member of the European Economic Community by the early 1990s. Despite periodic shortages of foreign exchange, Turkey is a goldmine for engineering contractors, with a string of infrastructure schemes from highway construction to vast hydro-electric power projects underway simultaneously.

Since the first suspension bridge crossing of the Bosporus was opened in October 1973, traffic demand for more crossings has increased dramatically. Within two years the number of vehicles using the bridge had risen from 28,000 to 52,000 a day. By June 1986 the bridge was choked with 100,000 vehicles a day. The capital cost of construction was recovered within three years from toll revenue [22].

A second bridge now under construction is predicted to become overloaded by 1994. Discussions to commission a third bridge within five years are well advanced with the British construction group Trafalgar House [23,24]. In partnership with the Turkish group Enka, it is hopeful of winning the $457 million contract after receiving a letter of intent from the Istanbul authorities in March 1987. The Trafalgar House proposal was chosen ahead of plans by the giant French construction group Bouygues for a $513 million road and rail tunnel under the Bosporus. Trafalgar House and its Cleveland Bridge subsidiary were beaten in the bidding for the second bridge by a consortium of Japanese, Turkish and Italian companies and controversy erupted over generous "soft loan" support from Japanese banks for the consortium which the British Government refused to match. The consortium put in a $729 million bid against $894 million from Trafalgar. The second bridge, a joint venture between Ishikawajima Harima Heavy Industries, Mitsubishi, Nippon Kokan, of Japan and Turkes-Feyzi Akkaya Insaat of Turkey and Impregilo of Italy, is on schedule for completion in October 1988. The dual four-lane, bridge five kilometres north of the existing crossing has a single main span of 1090 metres and 64 metres of clearance above the water.

The project includes 28 kilometres of approach motorway, linking it to the E5 motorway between Europe and Asia, known as the Kinali Sakarya road.

Hong Kong

The crowded island city has a network of tunnels to cope with the traffic needs of its teeming population of almost 5.5 million. The third harbour immersed tube tunnel crossing to link to Hong Kong Island with Kowloon is being built in a $450 million project by Japanese construction giant Kumagi Gumi. The combination road and rail tunnel of just under two kilometres is expected to be completed by 1989. It is one of the biggest construction projects in Asia. Kumagai Gumi built the tunnel crossing of the harbour on the Mass Transit Railway between Hong Kong Island and the New Territories. When it was opened in 1972 the 1.85 kilometre Cross-Harbour Tunnel was the longest submerged tube tunnel in the world. It has been such a success that it became the world's busiest four-lane road carrying 110,000 vehicles a day in 1983. The tolls ranging from HK$2 to HK$20 were kept unchanged but a tax introduced in 1984 initially cut usage by 15 per cent but it has crept back steadily to jam the tunnels with more than 100,000 vehicles a day [25].

United States

The world's largest immersed tube tunnel runs 2.7 kilometres between Baltimore and Fort McHenry in Maryland [26]. The tunnel, the final link in the US east coast Interstate Route I-95, carries eight lanes of road traffic. The contract for the tube fabrication and placement was the largest single

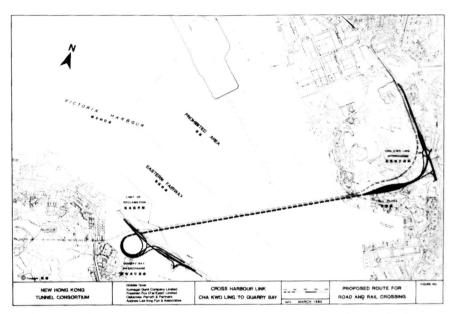

The eastern cross-harbour link in Hong Kong, the third tunnel across this stretch of water. Already a fourth one is planned.

project undertaken in the interstate highway programme. It was completed in November 1985, on time after five and a half years and $75 million under the proposed $8.25 billion budget.

Ports and shipping

British ports have been faced with the choice between adapting or dying for almost three decades. For the days when Britain was the hub of the world maritime trade, its docks crammed with liners and merchantmen, went forever with the passing of Empire and the arrival of containerisation.

As the opening of the Channel Tunnel looms, it is likely the intense rivalry that has characterised the development and decline of British ports will fade as the inevitability of cooperation to ensure survival becomes inescapable. The shipping lines and ferry companies must also reconcile themselves to fighting the threat together or sinking under the weight of their own intractability. That cooperation goes beyond the ports, shipping lines, distribution companies and ferry operators immediately challenged by the Tunnel and the development it is already spawning on the French side where the relationship between centrally planned investment supporting regional and local interests is already established.

In Britain the adversarial nature of society, filtering down from Parliament and the judicial system, can be seen clearly in rifts at the day to day level, the official indifference to the hardship of unemployment, the prosperous south's disdain for the industrially ravaged north and mutually defeating commercial rivalry between neighbouring regions, cities and towns. Only when consensus, a commitment for common benefit, the old Conservative ideal of "one nation" reasserts itself will Britain have the capacity to cope with the challenges the Tunnel, and the years ahead, will bring. The only alternative is to think "European", but that is the second stage of this enforced evolution. It will take the effects of the Tunnel's operation to finally bring Britain into Europe.

As Britain has turned itself slowly away from colonial markets towards Europe, now the source and destination for almost two-thirds of the nation's trade, the west coast maritime strongholds of Liverpool, Clydeside, Bristol and Cardiff have declined inexorably. Containerisation was a crucial factor, destroying thousands of stevedoring jobs as the work once done by hundreds was now a matter for handfuls operating sophisticated cranes and lifts. What was once done in days is now done in hours. It has been a vicious destructive cycle. And the ports of western Europe — Rotterdam, the largest container port in the world, and Antwerp rebuilt and modernised in the post-war regeneration and standing at least a day closer to the industrial complexes of Europe's heartland — were ready and waiting to capitalise on the unreliability of strike-happy British dockers.

The west coast ports bickering over the scraps of the Atlantic trade have bled almost to death. Lately their managements and workforces pausing in their battles with each other have found a new realism. Clydeside, where ships' numbers have fallen significantly over the years, Liverpool and Bristol are seeking Government support for a marketing push to attract new business from the trans-Atlantic trade on the basis that, because they are 30 hours closer to America than Rotterdam, with a direct rail link, a lot of expensive ship time could be saved. On the Clyde the Glasgow Confederation of British Industry (CBI) is calling the idea, "Eurowestport" [26,27]. Shipping is discovering along with other sectors of British industry that the only way to win and keep business is to perform — to a price. It was a lesson already learnt on the east coast. The east coast ports have thrived. Their growth, particularly of Felixstowe, has been the only encouraging side to Britain's port industry in recent years. Felixstowe has boomed and Harwich, Ipswich and Immingham have prospered [28]. It has not just been a matter of less militant workforces sparing operators the prolonged strikes that left Merseyside and the Clyde with a lingering taint to deter investors and entrepreneurs. Geography helped, as did the availability of relatively cheap commercial land and improved road links.

Felixstowe, operated by P&O Lines, is likely to maintain its position despite the effects of the Tunnel. It plans an expansion totalling £100 million, to confirm its hold on the booming container trade to the Scandinavian and north European ports [30]. Operations' director Mr Robin Macleod told a Parliamentary Committee that few customers were expected to reroute through the Tunnel because few cargo destinations and sources were closer to Dover than Felixstowe.

"The Channel Tunnel is unlikely to have a discernible effect on Felixstowe." he said in January 1986. Felixstowe has been targeted by several of the ferry companies who are looking to diversify away from the Channel routes; until 1987 these routes were the prized Blue Riband runs. P&O took over European Ferries at the end of 1986 paying £287 million for the Townsend Thoresen fleet and route structure and the port of Felixstowe. City observers said the move would give European Ferries a stronger financial base from which to fight the Tunnel threat [31–33].

The challenge to Dover, as Britain's premier ferry port, comes not only from the Tunnel but from other operators, particularly the Sally Line operating between Ramsgate and Dunkirk [34].. On the French side the Dunkirk Port Authority is building a new terminal on 24 hectares and upgrading the existing one in the meantime. At Ramsgate, the Sally Line itself is spending £12 million in new harbour facilities together with £7 million from the local Thanet District Council. The port will be able to accommodate up to five ferries at a time. Thanet learnt a bitter lesson with the demise of the hovercraft service from Ramsgate. Because it did not have a direct rail link, too few foot passengers used the facilitty to justify the service. The council was ready to help too late.

But despite the confidence of Felixstowe, the general threat to Britain's ports, already beset by overcapacity and under-use, is real. Maritime experts are agreed that renewed and committed investment is all that stands between the ports and decline into oblivion.

A report in January 1987 from Britain's Civil Engineering Economic Development Committee warns that without significant capital investment in British ports they will lose out to French, West German and Belgian rivals which have the benefit of "far sighted" Government investment [35]. Apart from continued investment in port facilities the report points to urgent road improvements needed to link ports to industrial areas, particularly the A1–M1 link between the Midlands and Felixstowe, completion of the A55 and the A5, completion of the M20, the Folkestone to Dover Road and the A2 near Dover.

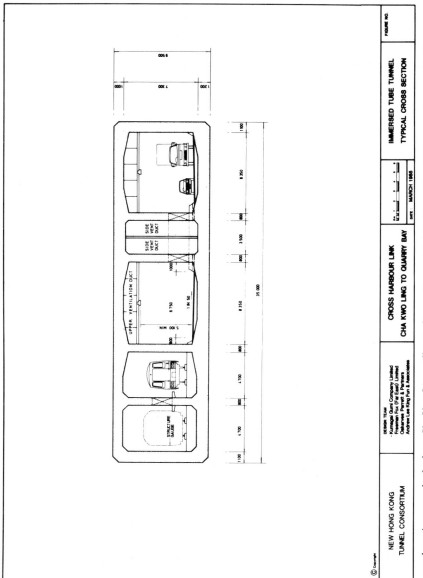

A very large immersed tube from Cha Kwo Ling will go to Quarry Bay carrying the Hong Kong Metro as well as road traffic.

The opening of the Channel Tunnel could eliminate two ports the size of Felixstowe, Sir Frederick Bolton chairman of the Association of British Ports said in April 1986. British ports faced being badly disadvataged because the Government refused to compensate those that suffered directly the effects of the Tunnel [36]. He cited the example of Le Havre, where it is expected a request for a motorway to be built from the port to the tunnel is likely to be agreed. At the same meeting Mr John Evelyn, chief executive of the Port of Ipswich, said Le Havre could become a virtual British port with the capacity to handle up to 40 per cent of British container trade before it was trans-shipped by road to enter Britain through the Tunnel. This, combined with the prospect that container freight trains from Britain could go through the Tunnel directly to Marseille, eliminating the need for ships from the Far East and Australasia to venture beyond Gibraltar, would mean the death of several ports, at the very least cause substantial losses to Southampton and Liverpool.

The occasion presented another opportunity for the annual plea from the British shipping industry for an alternative to the £50 million annual bill for light dues (lighthouses and pilotage) and £25 million for dredging costs, which are not borne by European competitors. If the UK Government concedes anything to the hard pressed maritime industry this is the most likely area. It is a small amount compared to what the operators themselves are spending and it would make a substantial difference to operating costs.

The industry is continuing to press the Government, so far with little effect, for measures to arrest the alarming decline in the merchant fleet. In March 1987 *Shipping World and Shipbuilder* reported the UK merchant fleet had shrunk to around 480 ships of about seven million tonnes total deadweight, a mere 1 per cent of the world fleet. In 1975 UK flag merchant vessels amounted to 50 million tonnes deadweight and in 1980 the figure was 35.7 million tonnes deadweight. In the same report the General Council of British Shipping said the 1985 Treasury forecast of between five and seven million tonnes deadweight by 1990 is "wildly optimistic". Orders for new ships had virtually dried up, declining from 363,000 tonnes deadweight in 1983 to 258,000 tonnes deadweight in 1985 and 5000 tonnes deadweight in the first half of 1986. According to a prediction made to the Steam Ship Association in March 1986 by Sir Brian Shaw, then president of the General Council of British Shipping, the UK fleet could fall to only 200 ships totalling five million tonnes deadweight by 1995. He compared this with a forecast ten years earlier which said the 1995 figure would be 1600 ships of 50 million tonnes. He called for deferred tax liability as an incentive for shipowners to further investment. So far the Government has not responded [37].

The strategic consequences for Britain in the decline of the merchant fleet, exacerbated by the Channel Tunnel effects on the ferries, has been a plank in the opposition mounted by Flexilink, the anti-Tunnel campaign heavily backed by the ferry companies. Quoting the Sea Group report compiled by Lloyds on the contribution of the merchant fleet to Britain's defence needs, it said about 80 ferries would be needed for military supply duties in the event of conflict. About 100 of the 114 ferries around the UK are suitable for military use and Flexilink argued that any reduction in the ferry fleet so far unaffected by the decline in merchant tonnage would potentially leave it inadequate to respond to military needs. Although this did not take into account the value of the Tunnel as a direct supply route across the Channel its opponents argued that should it be subjected to concentrated attack, it would be made unusable for either reinforcement of forward forces in Europe or withdrawal back to the UK. It is an argument that bears consideration. There are still many who remember how the Dunkirk retreat was possible only because of the flexible capability provided by "many ships and many ports". It is an idea fundamental to the belief that "Little England" stands alone against its enemies [38,39].

While waging a relentless campaign of opposition against Eurotunnel, Sealink has been making preparations against the inevitable with suggestions of an amalgamation with Townsend Thoresen on the prized Dover to Calais route. Its opening gambit of a half-hourly shuttle that might use a few as five ships was met with scepticism from P&O owned Townsend Thoresen. But like Sealink it wanted the Government to lift its cartel rules banning price cooperation between the ferry operators, so they could combat the Tunnel with a joint strategy. It has been expressly forbidden under monopoly regulations but when the Tunnel becomes operational it would be surprising if the Government did not grant some concession. Once the Tunnel is open then considerations of monopoly exploitation become irrelevant. With steadily rising demand, ferries can expect good profits until the Tunnel arrives. The Irish Sea connections continue to be troubled and the Channel Island services run by Sealink are also a steady loser. No one can be certain of the size of the expanding market that will be left for the ferries after the Tunnel opens.

The UK National Union of Seamen predicted that the Tunnel would wipe out most of the 34 UK-flag ferries operating between Britain and the Continent with about four cargo-only vessels surviving on the routes between Hull and Plymouth. The union told the Commons Select Committee this would mean redundancy for 5600 seamen employed in Dover, Folkestone, Hull, Harwich, Felixstowe and Plymouth. The ships

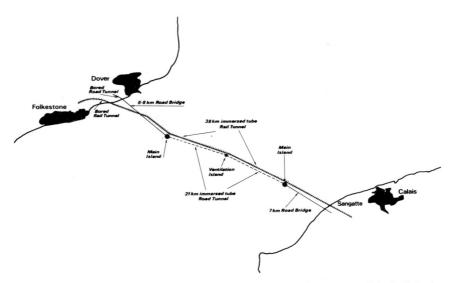

EuroRoute's cross-Channel island, tunnel, bridge, scheme may still be built in the future... or something along similar lines will be designed.

on the larger routes between Hull and Harwich to the Hook of Holland were most likely to survive but this would depend on cooperation between the fixed-link operators that would effectively close Dover to cut-price flag-of-convenience operators. Such and arrangement would mean the survival of about 2,000 seamen's jobs [40].

The ferry companies have reacted to the Tunnel agreement with anger and a determination to meet the challenge head-on with heavy investment in ships and facilities. Townsend Thoresen have already taken delivery of the first of two 20,000 tonne "jumbo" ferries from West German shipbuilders, ordered for £85 million. Twice the size of vessels previously used, each can carry 120 trucks or 700 cars and 2400 passengers. Plans for a third vessel have been put to one side. Sealink delayed decisions on new ships for the Strait of Dover crossing but has set up ventures in the Mediterranean and examined routes to Scandinavia and the Iberian Peninsula in an attempt to diversify away from dependence on the cross-Channel market.

The go-ahead for the Channel Tunnel presents the port of Dover with the biggest threat it has faced since it was established under Royal Charter in 1606. It is not under threat of extinction; it has survived as a port for 2000 years and it is the closest UK port to the European mainland. In his report for 1985, Dover Harbour Board chairman Sir Frederick Bolton said: "The port shall be geared to live with the future as its shape emerges. The Tunnel decision anyway has all the appearance of being political, for the economic evidence against it seems almost entirely undigested." Dover's response to the latest Tunnel threat has been a frenzied investment programme, dwarfing the steady improvements it has made in the past 30 years. It plans to improve facilities to cater for the new generation of super ferries, to reclaim land and upgrade freight handling. Traffic levels through the port have increased sharply since 1976 with the number of vessels up from 19,591 in 1976 to consistently more than 28,000 a year since 1984. In 1986 it handled more than 14.3 million passengers, an increase of almost 600,000 on 1985 and more than double the 1976 figure, many of them on holiday coaches whose number quadrupled in the ten years. Commercial freight lorries almost trebled in the same period, to 858,774 [41].

The opening of the Channel Tunnel could be the impetus for a second generation of high-speed hovercraft ferries capable of taking both coaches and cargo over greater distances than the present machines which have been in service since 1970, according to Hoverspeed which only operates on the Dover to Calais route. Hoverspeed's operations' manager, Mr Derek Meredith is looking to diversify to other, longer routes, including the stormy Irish Sea to beat the Tunnel's direct onslaught on its market. The new craft would carry 600 or more passengers and up to 80 cars compared to the 424 people and 55 cars currently thought possible. They would be able to travel at 80 knots, ten knots faster than the present craft. On the Portsmouth–Le Havre route advanced hovercraft could cut three and a half hours off the ferry sailing time. Felixstowe or Harwich to Rotterdam or Zeebrugge have also been cited as possibilities. But the fear remains that if Eurotunnel achieves its target of 70 per cent of the cross-Channel market the remaining 30 per cent will be insufficient to justify the ferry operators maintaining services at a number of other ports apart from Dover, namely Folkestone and Ramsgate. Job losses are a sensitive issue and the debate rages around the question more fiercely than any other. Nicholas Finney, director of the British Ports Association. claims that ultimately some 176,000 jobs will be lost to the Tunnel. The Transport and General Workers Union predicts 50,000 jobs at immediate risk. The Kent Impact Study envisages port-related job losses to be up to 6000. Across the Channel, Calais handled more than nine million passengers in 1986, an increase of some 12 per cent on 1985. Calais, like Dover will be by-passed by the Tunnel and will suffer job losses too. Yet while the ferry traffic is a vital part of the French ports plans it is preparing to catch far larger fish than the new 20,000 tonne super-ferries. A long-planned deep water dock capable of handling ships of up to 45,000 tonnes is planned for completion in 1989 and port authorities have signalled their intentions with predictions that Calais "will enter a new era of industrial life". They have already begun inviting international investment, particularly from

Britain. Boulogne too is making serious preparations for the arrival of Tunnel traffic. The Garromanche project, a joint enterprise between the Boulogne Council and cold store and freight distribution operators, is geared to integrate road and rail services specifically handling British imports and exports. The temperature controlled warehousing will provide a launch pad for French produce exports.

The French Government has committed itself to spending almost £600 million on connecting Calais with the national motorway network on a road that will link all the Channel ports from Dunkerque to Le Havre. It has promised a further £30 million on upgrading rail links to Calais. The French have also moved fast to try ensure that the lion's share of jobs created by the Tunnel go to northern France. They have made no secret of their intention to make the Nord Pas de Calais the crossroad of Europe. Their traditional determination to stimulate job growth with grants and cash incentives has been surpassed by a programme which will:

• Grant FFr50,000 (£4761) for each job for the creation of 20 jobs in three years

• Give small businesses FFr10,000 (£952) for creating three jobs for three years

• Pay a firm creation grant of FFr50,000 for three jobs, FFr100,000 for six jobs and FFr150,000 for ten jobs or more.

• Offer building loans for 75 per cent of cost, at a fixed interest rate of 8.75 per cent.

• Subsidise 15 per cent of computer equipment purchases.

On the British side there are none of these incentives. While the French forge ahead with development plans specifically aimed at exploiting the Tunnel trade route it will be mid-1988 before the British consider the details of comparisons between financial support systems in Kent and northern France. The long-term employment benefits will be almost exclusively French and what could have been a once in a lifetime opportunity to significantly alter the decline of the past two decades will be lost. Professor Paul Ormerod, of the Henley Centre and the University of London, has started to research the effects of the Tunnel on business with particular regard to the northern UK as the survey is backed by The Association of Greater Manchester Authorities. Early results show that most businessmen expect the greater part of all benefits to go to Kent and the south-east of England where unstoppable urban sprawl is certain to roll down the A2 corridor into Kent as more and more people flock to the work [42]. There is no guarantee that any action will be taken to correct any imbalance and unfair advantages uncovered, as the British Government sticks to its commitment of not subsidising the Tunnel with public money with the tenacity of a martyr trying to die for his creed. Unless

The Sally Line is one of many shipping lines that will be affected by the Tunnel and it will
have to adapt and evolve rapidly to survive. The Viking sails out of Ramsgate on the two-
and-a-half-hour journey to Dunkerque in France.

Britain backs up its political will to support the Tunnel with financial
muscle. The predictions that the Channel Tunnel will become a one-way
drain on prosperity will be fulfilled.

Into the Future

Tunnels exert a remarkable lure on the imagination. Perhaps it is the
mystique of going into the dark heart of the earth, or is the wonder of
disappearing like Alice in Wonderland and emerging at one's destination, having travelled unfettered and unimpeded from point to point, an
impossibility on the surface? The reasons matter little, it is sufficient that
advocates believe nothing is impossible.

Hugh Burn, a 66-year-old local authority chartered engineer now living
in quiet retirement in Hampshire, has an idea for a Channel Tunnel that
makes anything proposed so far seem timid and archaic.

Since 1965 he has nurtured a dream of the future which would take a
single bore tunnel from the heart of London's underground "Tube"
system straight to a central connection with the Paris Metro, at Chatelet.

The journey, city to city, would take just 25 minutes as an automatically-guided vehicle carried up to 1000 passengers as fast as a bullet, 1000 kilometres an hour, through the Tunnel stretching 345 kilometres without a bend.

Hugh Burn believes that this, or something like it, will happen one day. "I don't really expect to see it happen in my lifetime, now," he said as the House of Lords Select Committee was sitting in judgement on the Channel Tunnel. "It can be done. If people applied their minds to it, my idea for a high speed London–Paris tunnel could be operating by the late-1990s."

Of course, he is right. The scope for development of new transportation techniques is unlimited as technology gallops continually further ahead of human ability to imagine its potential.

Beyond the turn of the century it could be an integral part of City business life to enjoy a superior working lunch in Paris knowing that the hourly "Tunnel Shuttle" makes it possible to be back at the desk by the end of the day with time to check the closing prices and sign letters, before taking the evening commuter train to the suburbs an hour away. "Taking the Tunnel" to Europe in less time than it takes to rattle to Heathrow by taxi, or underground, would change concepts of distance and mobility just as dramatically as the aeroplane. London theatres, Parisian restaurants and clubs, the great museums and galleries of both cities would be virtual neighbours.

Mr Burn has chosen the centres of Paris and London as his termini because they are the centre of greatest demand. Equally it could be between any other major centres. Paris and London are already equipped with sophisticated networks geared to bringing passengers into the national hubs and to disperse them once they have arrived. Burn visualises a service dedicated to passengers. He admits the lack of freight is a commercial drawback but with some relief points to the increased margin of safety this offers users. He believes that trying to handle vehicles and freight in city centres already overburdened with traffic would create insurmountable problems. Drivers with journey origins and destinations at widely differing points would be better served by an upgraded ferry service.

The selling point of the proposal is the remarkable speed. While the idea of travelling underground at 1000 kilometres an hour, the cruising speed of modern jetliners, may seem too fantastic to consider seriously, Mr Burn maintains it is a modest goal. He refers to an idea put forward by Dr Robert M Salter, of the Rand Corporation of America (The Very High Speed Transit, Rand paper p.4875, August 1972) in which Dr Salter forecasts worldwide underground trains travelling at 22,400 kilometres an hour by electrically levitated and propelled cars in a tunnel."

"Of course," writes Burn, "This is rather like trying to run before we can walk but it does suggest 1,000 an hour is a fairly modest target."

In a comprehensive paper detailing his proposals Burn said: "the vehicle would be propelled by a series of static electric coils wound on an open framework close to the vehicle for maximum electro-magnetic influence."

"The vehicle would act as the central core passing through a series of solenoids. As it is attracted towards the centre of the magnetic field of each core the current is automatically switched to the next coil, then to the next and so on. The coils will installed at spacings which increase according to their distance from the terminus to give a controlled and comfortable rate of acceleration and, ultimately deceleration. Advantage would be taken of the momentum of the vehicle, which at maximum speed would be impelled from one coil to the next with minimal energy loss, compared to traditional traction.

Burn describes a vehicle 300 metres long capable of carrying 1000 passengers in hermetically-sealed comfort, seated four abreast, back to back.

While propulsion is electromagnetic, Burn believes that the expense of magnetic levitation outweighs its benefits. Aerofoils under the vehicle floor would give lift, thereby reducing drag on the wheels, much in the way a hydrofoil hull can be lifted clear of the water and friction reduced to a minimum. A top rail guide could prevent the machine drifting too far off course.

Burn proposes the machine should be cylindrical, about four metres in diameter. Travelling concentrically in the seven-metre-diameter tunnel, the shuttle would leave an annulus of around one and a half metres for air displacement, sufficient to cope with any proplems with overheating or wind resistance. Rapid transit with one vehicle each way every 30 minutes would avoid ventilation problems. The plan incorporates a simple but effective passenger management system — each terminus excavated deep below existing underground stations would incorporate platforms on either side: one for passengers boarding and one for those alighting.

Burn priced his scheme at about three times the cost of the Channel Tunnel proposals, amounting to close to £20 billion in April 1987. He estimates that a shuttle could operate 16 hours a day, carrying 1800 passengers at £150 a journey or a price to compete directly with aircraft services. The system would have a lifetime of 120 years. "This could be the fastest, safest, most reliable transport system in the world," said Burn, pleased and proud his plans had been noticed. "Nobody has rejected the idea itself, they just say it is too expensive. "This is the logical step beyond a coast to coast fixed link. "It seems to me that people are not looking far enough ahead."

References

[1] Missing Links: Upgrading Transborder Ground Transport, Round-table of European Industrialists.
[2] *The Times*, Jan 10, 1983.
[3] *The Times*, Jan 21, 1986.
[4] *The Times*, Dec 10, 1984.
[5] Lloyds List, Jun 14, 1986.
[6] Lloyds List, Mar 20, 1986.
[7] *The Times*, Aug 7, 1986.
[8] *La Vie Economique*, Feb 7, 1986.
[9] The *Observer*, Nov 9, 1986.
[10] *The Messina Strait Crossing*, Stretto di Messina 1981.
[11] *International Herald Tribune*, Mar 26, 1986.
[12] *Engineering News Review*, Apr 3, 1986.
[13] *Daily Telegraph*, Jan 3, 1986.
[14] *International Herald Tribune*, Jul 3, 1986.
[15] *The Times*, Mar 11, 1985.
[16] *The Guardian*, 13 Mar 13, 1985.
[17] *New Scientist*, Apr 25, 1985.
[18] *Civil Engineering in Japan 1985*, Japanese Society of Civil Engineers.
[19] *Seikan Undersea Tunnel*, Japan Railway Construction Public Company.
[20] The *Guardian*, Mar 13, 1985.
[21] *Tunnels and Tunnelling*, Mar 1986.
[22] *World Construction*, Jun 1986.
[23] *Financial Times*, Mar 7,1987.
[24] *New Civil Engineer*, Dec 11, 1986.
[25] *1986 Hong Kong Yearbook: Tunnels and Tunnelling*, Jul 1971.
[26] *Scotsman*, Jul 3, 1986.
[27] *Daily Telegraph*, Jul 3, 1986.
[28] Lloyds List, Dec 31, 1986.
[29] *Export Times*, Sept 1986.
[30] Lloyds List, May 12, 1986.
[31] *Travel News*, Jul 23, 1986.
[32] *Financial Times*, Oct 22, 1985.
[33] *Daily Telegraph*, Mar 31, 1986.
[34] Civil Engineering Development Committee, Infrastructure Report Jan 1987.
[35] *The Times*, Oct 17, 1986.

[36] *Financial Times*, Oct 22, 1985.
[37] *Shipping World and Shipbuilder*, Mar 1987.
[38] *Janes Defence Weekly*, Mar 1987.
[39] Lloyds List, Jul 30, 1986.
[40] Lords Hansard, Apr 2, 1987.
[41] Dover Harbour Board, 1986 Annual Report.
[42] *The Channel Tunnel and the North of England*, Henley Centre, Apr 198.7

Index